Transforming Futures: The Brooklyn Program Facilitator's Manual

Second Edition

Richard M. Gray

Copyright 2008-2012

by Richard M. Gray, Ph.D.

All Rights Reserved

Library of Congress-cataloging in-publication-data

Gray, Richard M.

Transforming Futures: The Brooklyn Program Facilitator's Manual / Richard Gray.—second edition.

Includes bibliographical references.

ISBN 978-1-257-96354-6

1. Addiction and recovery. 2. Psychology. 3. Neuro-Linguistic Programming.

Contents

INTRODUCTORY MATERIALS i

A Letter to the Reader i

    Presuppositions ii

    Tools iii

    The Program iii

Forward vii

    APPLICATIONS OF THE PROGRAM. IX

    ORGANIZATION OF THE TEXT XI

Acknowledgments xiii

Introduction: Thinking About Drugs and Addiction 1

    THE THING WE CALL ADDICTION 2

    NEUROSCIENCE AND ADDICTION 4

    NLP APPROACHES TO ADDICTION 7

    REFERENCES 12

Basic Competencies: Exercises 1-7 17

Introduction to Exercises 1-7: Finding States, Anchoring and Future Pacing 17

The Exercises 25

Exercise 1 Feeling Good 25

    PRESUPPOSITIONS UNDERLYING THE EXERCISE: 25

    EXPECTED OUTCOME 26

    INSTRUCTIONAL NOTES 27

Process Summary and Script *31*

BEHAVIORAL STANDARDS 35

SEQUENCE NOTE: 35

EXERCISE 1 FEELING GOOD 37

Further Applications *43*

References for the exercise *43*

Exercise 2 Finding Resource States 45

PRESUPPOSITIONS UNDERLYING THE EXERCISE: 45

EXPECTED OUTCOME 46

INSTRUCTIONAL NOTES 46

SEQUENCE NOTE: 52

BEHAVIORAL STANDARDS 53

EXERCISE 2 FINDING RESOURCE STATES 54

Further Applications *56*

References for the exercise *57*

Anchoring Resource States 59

PRESUPPOSITIONS UNDERLYING THE EXERCISE 59

EXPECTED OUTCOME 61

INSTRUCTIONAL NOTES 61

BEHAVIORAL STANDARDS 65

MEDITATION 66

THE ANCHORING EXERCISE PROVIDES A PROPER END TO THE SESSION 66

Exercise 3 Anchoring Resource States 67

Creating Anchors *68*

The basic states *70*

Further Applications *71*

References for the exercise *71*

Exercise 4 Keys to Enhancing Subjective Experience 73

PRESUPPOSITIONS UNDERLYING THE EXERCISE: 73

EXPECTED OUTCOME 74

INSTRUCTIONAL NOTES 74

BEHAVIORAL STANDARDS 74

MEDITATION 74

Initial Strategies *75*

Further Applications *77*

References for the exercise *77*

Exercise 5 Getting to NOW 79

PRESUPPOSITIONS UNDERLYING THE EXERCISE: 79

EXPECTED OUTCOME 80

INSTRUCTIONAL NOTES 80

Script for creating NOW *80*

BEHAVIORAL STANDARDS 83

MEDITATION 83

EXERCISE 5 GETTING TO NOW 84

Getting to NOW *84*

Further Applications *85*

References for the exercise *85*

Exercise 6 Pacing The Future 87

PRESUPPOSITIONS UNDERLYING THE EXERCISE: 87

EXPECTED OUTCOME 88

INSTRUCTIONAL NOTES 88

BEHAVIORAL STANDARDS 89

MEDITATION 89

EXERCISE 6 PACING THE FUTURE 90

Further Applications 93

References for the exercise 93

Exercise 7 Taking Control 95

PRESUPPOSITIONS UNDERLYING THE EXERCISE: 95

EXPECTED OUTCOME 95

INSTRUCTIONAL NOTES 96

BEHAVIORAL STANDARDS 96

MEDITATION 96

EXERCISE 7 TAKING CONTROL 97

The Exercise 97

Further Applications 98

References for the exercise 98

Developing Self and A Sense of the Future: Exercises 8-12 101

Introduction to Exercises 8-12: Positive Resources, Setting Goals, ImageStreaming, Sponsoring a Potential 101

Exercise 8 Positive Experiences 105

PRESUPPOSITIONS UNDERLYING THE EXERCISE: 105

EXPECTED OUTCOME 106

INSTRUCTIONAL NOTES 106

BEHAVIORAL STANDARDS 108

MEDITATION 108

EXERCISE 8 POSITIVE EXPERIENCES 109

    Instructions *109*

    Further Applications *110*

    References for the exercise *110*

Exercise 9 Positive Experiences Revisited **113**

  PRESUPPOSITIONS UNDERLYING THE EXERCISE: 113

  EXPECTED OUTCOME 114

  INSTRUCTIONAL NOTES 114

  BEHAVIORAL STANDARDS 115

  MEDITATION 115

  EXERCISE 9 POSITIVE EXPERIENCES REVISITED 116

    Instructions *116*

    Further Applications *117*

    References for the exercise *117*

Exercise 10 Setting Goals That Work. **119**

  PRESUPPOSITIONS UNDERLYING THE EXERCISE: 119

  EXPECTED OUTCOME 121

  INSTRUCTIONAL NOTES 121

  BEHAVIORAL STANDARDS 127

  TRANCE / MEDITATION 127

  EXERCISE 10 SETTING GOALS THAT WORK 128

    Instructions *128*

    Outcome Worksheet *133*

    Further Applications *134*

    References for the exercise *134*

Exercise 11 ImageStreaming Into The Future **137**

- PRESUPPOSITIONS UNDERLYING THE EXERCISE: 137
- EXPECTED OUTCOME 138
- INSTRUCTIONAL NOTES 138
- BEHAVIORAL STANDARDS 139
- MEDITATION 139
- EXERCISE 11 IMAGE STREAMING INTO THE FUTURE 140
  - *ImageStreaming into the Future* *140*
  - *Further Applications* *142*
  - *References for this exercise* *142*

Exercise 12 Sponsoring a Potential 143

- PRESUPPOSITIONS UNDERLYING THE EXERCISE: 143
- EXPECTED OUTCOME 144
- INSTRUCTIONAL NOTES 144
- BEHAVIORAL STANDARDS 146
- MEDITATION 146
- EXERCISE 12 SPONSORING A POTENTIAL 147
  - *References for the exercise* *148*

Scripts 151

- SCRIPT ONE ENHANCING RESOURCE STATES 152
- SCRIPT TWO ANCHORING A RESOURCE STATE 157
- SCRIPT THREE CREATING NOW 160

Meditations 163

- INTRODUCTION 163
- REFERENCES 164

Meditations New Worlds to Gain 165

| PRESUPPOSITIONS UNDERLYING THE EXERCISE: | 165 |
| INSTRUCTIONAL NOTES | 165 |

Meditations Future Perfect 169

PRESUPPOSITIONS UNDERLYING THE EXERCISE:	169
INSTRUCTIONAL NOTES	169
MEDITATIONS FUTURE PERFECT	169

Application Notes 175

Application Note 1 177

Application Note 2 181

Application Note 3 185

Application Note 4 191

Application Note 5 195

The One-on-One session *195*

Application Note 6 201

Application Note 7 205

Application Note 8 207

Application Note 9 213

General Session Procedures 217

| PROGRAM TIMELINE | 218 |

The One-On-One Sessions 221

BEHAVIORAL CRITERIA	221
THE POSITIVE RESOURCE DAY PLANNER	225
POSITIVE RESOURCE DAY PLANNER	226

The Final Session 227

Appendix Forms and Handouts 229

Transforming Futures

- POSITIVE RESOURCE DAY PLANNER 231
- OUTCOME WORKSHEET 233
- BEHAVIORAL CRITERIA 235
- EVALUATION 237

References 239

INDEX 245

INTRODUCTORY MATERIALS

A Letter to the Reader

Several people have asked about the application of the Brooklyn Program to problems other than substance use disorders. I believe, based on the responses of my own offender populations, that it can be useful for addressing general criminal behavior and sex offenders as well. Moreover, whenever I have taught it to non-criminal populations, their response has been overwhelmingly positive. While not a panacea by any means, it does provide the kinds of experiences and shifts in personal direction that can awaken movement towards change in almost any population.

Until my retirement in 2004, I worked for almost thirty years with state and federal probationers and parolees. During the last ten years of that period I developed the Program based on the technology of Neuro-Linguistic Programming (NLP) for a range of offenders who had been determined to suffer from substance use problems —including those who were assessed as having a problem simply by virtue of being in the same room as drugs or talking about them on the phone—as well as persons with real drug problems. The Program was rooted in Jungian and Maslowian presuppositions about the nature of human development and used NLP tools to create a multilayered series of changes. An important part of the Program was the structuring of long term goals that were consistent with a sense of calling or deep personal direction. Beyond the Jungian and Maslowian elements this piece was modeled on the findings surrounding the Stages of Change (or the Trans Theoretical) Model developed by Prochaska, DiClementi and Norcross.

The Stages of Change Model is currently the most well-researched and validated pattern for understanding how people change. It indicates that people go through five discrete stages when they are changing—whether for the better or for the worse. The stages are: precontemplation—when the subject may not even be aware that a change is necessary or desirable, contemplation—when the subject begins to consider the possibility that change could be good, preparation—where a decision has been made that some kind of change is necessary and should begin in a short time, action—when the changer is committed to a plan of action, and maintenance—when the changer no longer identifies with the problem behavior. A sixth stage, termination reflects the time when the changer gets on with life without regard to the problem.

One of the important predictors of movement through the changes is the identification of a positive future that is incompatible with the problem behavior. NLPers know about this in terms of well-formedness conditions for outcomes. See below.

The Program lasted for 16 weeks with weekly sessions of two hours each. More than 300 people passed through the Program over a period of seven years. There are several articles about the Program on my website at http://richardmgray.home.comcast.net

After using the Program for several years I discovered that it created such a radical reorientation in most people that it could be used as a general, whole-life reframe without regard to the problem under consideration. I have found that wherever it was taught, especially to service providers, their lives were changed as much as the offenders for whom it was originally designed.

Presuppositions

There are several presuppositions that undergird the Program. Most of them are standard parts of the NLP world-view. Some of the most important are these:

- Most people already have the resources that they need.
- People are, for the most part not broken.
- All behaviors can be analyzed into their sensory components (what is seen, heard, felt, smelled and tasted) and by reassembling those components, we can recreate the behavior or perception in another person or in the same person at another time.
- Further, by examining the submodalities, the finer grained structure, of each of those senses—like brightness, volume, timbre, color, warmth, movement—we can change the meaning or intensity of the experience.
- Finally, for the most part, choice is better than no choice.

The Program depends heavily on the presuppositions of general systems theory. The most important presupposition from this source is that complex systems often give rise to emergent properties that could not be predicted from the properties of the original elements. Simply stated, when things are combined in new ways you often get results that you may not have expected.

One presupposition that is specific to this Program is that there are motivations and outcomes that are inherently transformative. These transformative outcomes, though possibly universal as to types, are unique to each individual.

Tools

The Program uses several basic tools from the classic NLP toolbox. The first is Submodality analysis and manipulation. Simply, this means that by noting and adjusting the fine structure of an experience we can change its power and meaning. Notice the difference that you experience when thinking about a pleasant memory; first, as if it were a distant, black and white memory, projected on a movie screen and then, as a bright, three-dimensional presence, experienced from within your own body, with all of the smells, colors, tastes and feelings that are a part of the experience. Let it wrap around you and notice how things move. These are some of the differences that submodalities make.

The second tool is called anchoring. In NLP, anchoring can refer to almost anything from a gentle touch used as a conscious reminder, to a classically conditioned stimulus that evokes a specific, involuntary, emotional or visceral response. In the Brooklyn Program, anchoring is treated as a classically conditioned learning experience in which repeated pairings of a meaningless gesture with an emotional experience allow that gesture to elicit and modify the original emotional experience. These conditioned stimuli may be thought of as triggers for the desired response. They are automatic and relatively immediate. Stacking anchors, an extension of anchoring as an instance of classical conditioning, allows the creation of new states and experiences by combining felt experiences.

A third tool consists of the criteria for well-formed outcomes. This is a list of characteristics that meaningful and motivating outcomes have in common. It is rooted in the linguistic understanding of syntax that forms some of the basic concepts of NLP and matches the criteria of intrinsic motivators as identified by classical research. It includes tests like the following: Is the outcome stated in the positive—what you want or will have, as contrasted with what you don't want or will no longer have? Is it under your control—is it something that you could do for yourself? Can you specify how you will know that you have it when you get it? That is, can you tell me, in terms of what you will see, hear, feel, taste, and smell, what will be different, what will be there?

The Program

The Program begins by turning away from focusing on the problem and emphasizes that the participants can learn to enhance their memories, feel better emotionally, gain control over their emotions—choose how and when they want to feel differently, and finally, design a future that is meaningful to them. Problems are deemphasized. In some cases the Program was presented as laying a behavioral foundation for later work on the problem behaviors themselves.

In the first several sessions, participants are taught how to access and enhance a positive resource state using standard NLP submodality techniques. As any NLPer knows, this submodality work begins with a striking enhancement of the remembered experience and so validates the first promise that participants will learn to enhance their memories. During the same several sessions, the participants are taught to focus more and more on the feelings associated with the experience so that they discover a series of deeply-pleasurable transcendent states. These pseudo-meditative states are designed partly to provide feelings of self-efficacy, but also to provide powerful positive experiences that are strong enough to challenge the salience of the problem state.

Next, in sequence, the participants are taught to anchor several predefined states that they have accessed and enhanced. These include the experience of focused attention, a single good decision made in a systematic fashion, a moment of skill consolidation or streamlining of a learned behavior—riding a bike, driving a stick shift, an experience of pure fun or enjoyment, and an experience of confidence or personal competence. These resources are enhanced to ecstatic levels—to the point where there is virtually no shadow of the original content or context. Each state is anchored to a distinct hand gesture. The anchors serve three purposes:

- They make the resource transportable and accessible in multiple contexts,
- They create a relatively mechanical means for evoking and enhancing the anchored state,
- They create an automated access for later integration of these anchors into a more complex state (stacking anchors).

These five exemplars and the first level of stacked anchors were inspired by a set of anchors described by Carmine Baffa and were originally added to facilitate the later exercises.

Once the anchors have been practiced and enhanced several times. Participants are encouraged to practice them in multiple situations so that they generalize into other life contexts. They are also encouraged to create several of their own anchors to make sure that they understand that all of this is under their control and that they represent their own skills.

At about the seventh week, the anchors are stacked into a single anchor which I call NOW and which, according to my understanding creates a basic felt experience (constellation) of Jung's deep Self. This is important because it will provide an affective basis for creating a truly meaningful and compelling set of outcomes when in the last sessions we use the NLP well-formedness conditions to create a future that fulfills Prochaska's idea that movement through the stages of change is propelled most significantly by the identification of a meaningful and compelling future.

The process continues with the collection of another series of resources from various time periods in the participants' lives. These consist of times when the participants felt good about themselves, things that they did well, things that they learned easily, meaningful jobs and roles that they held, and things they wanted to be when they were kids. Although the original program called for these states to be anchored and integrated into the NOW anchor, I believe that their evocation using NOW as a window into the past, provides a sufficient level of integration.

Finally, the felt state associated with NOW is used to create smart outcomes across several life domains: home life, occupation, spiritual life, relationships, intellectual life, and health practices. Each outcome is created by accessing the NOW anchor and imagining life dominated by the felt state, "NOW," through the window of each of these domains. This results in future outcomes that are consistent with a deep, felt sense of personal identity. Superficial outcomes—wealth, sex, possessions etc. are discarded in favor of behavioral outcomes that characterize the kinds of behaviors that give expression to the constellated sense of the deep Self. The remaining exercises are devoted to enhancing the vision of the future and consolidating the learnings.

One of the more striking outcomes in the course of the Program was the near universal and spontaneous use of the anchors for anger management. It seemed that as soon as the participants found out that they had a reliable means to control their emotions, they began to use the anchors to create choice about how they were feeling in the moment. This is all the more striking in light of our commitment never to tell the participants how or where to use the anchors.

Many of the participants have told me that they wish that they had experienced the Program early in their corrections careers, whether inside a prison facility or while on the street, serving a term of community corrections.

Our results with substance abusers and addicts—persons with known histories of substance use and current use verified by urinalysis, ran to 29.6% abstinence after a year. I am sure that similar, positive results would apply for persons without substance use problems.

Forward

This Program developed over more than ten years of research into Jungian patterns, Neuro-Linguistic Programming, addictions studies and personal experience with addictions and an addictions caseload.

It is founded on the presupposition that people are, for the most part, whole and healthy and that an essential element of overcoming substance use disorders is the realization that we all have inner resources that can meet our present and future needs.

An important conceptual foundation of the Program is that **humans are systems**. We are integrated wholes who grow and develop in an orderly, systematic fashion. The elements of systems theory were first worked out by the biologist, Ludwig von Bertalanffy and have since been expressed in many other fields. Here, we apply systems presuppositions to the field of addictions and substance use disorders more generally. Most importantly, we predict that the systems property of *wholeness*, the emergent property of the proper interaction of all of the subsystems, can be manipulated by positive learning strategies. For our purposes, this means that **people can be taught positive skills that will interact with each other in such a manner that their whole way of experiencing life can be transformed.** It is through this transformation of experience that we allow the individual to move beyond addictive patterns.

This is an indirect approach. Most of today's standard answers to addictions take a head-on, allopathic stance that assumes that the addiction exists as a thing in and of itself and that by dealing with the issue, the urge, the triggers, the substance or chemical itself, we can solve the problem. By contrast, our approach holds that **addiction and substance use disorders more generally are expressions of systemic wholeness that answer very specific behavioral and emotional needs.** For each person they represent a specific need for integration and balance for which no other answer is currently available. Addiction, as a single phenomenon, does not exist, except as a behavioral context.

In this Program we seek to redefine the root structures of experience in such a way that new options are not only available, but desirable.

There are certain affinities between this approach and the older, psychodynamic *disease model* in which addiction was viewed as a symptom of some other, more fundamental deficit or disease. In that model, elimination of the symptom — addiction — would necessarily result in symptom substitution. It would follow from this perspective that the root problem must seek new expression in another object or behavior. By contrast, we see substance use disorders not as symptoms but, *given the resources available,* as the *best possible answer for the problems at hand.* Here, the re-emergence of symptoms speaks less of underlying pathology than it does of the need for a more fundamental restructuring of the available resources. By this definition, ***addicts and other substance abusers are not broken; they have simply learned the wrong answers to the questions of life***.

Our task becomes this: making proper answers available, making them more intuitive and more powerfully motivating than the focus of the addictive behavior.

There are three elements of treatment that seem to reappear with great consistency throughout the literature. These are *Self-Efficacy, Futurity* and *Self-Esteem*.

Self-Efficacy, originally defined by Albert Bandura (1997), describes *the individual's ability to experience a degree of control about themselves and their environment.* Addicts and substance abusers have experienced considerable and increasing control deficits in their lives. By definition, addiction entails a narrowing of focus and loss of control to the addictive behavior.

By restoring a sense of control, choice and mastery, this Program seeks to open options for new behaviors that are not focused in addictive patterns of seeking and using the addictive substance.

Futurity is held out by Prochaska, Norcross and DiClemente (1994) as a crucial predictor of treatment success. They point to the necessary shift from focusing on the loss suffered by giving up addictive behaviors to the perception of all that can be gained by moving on to a better future as the telltale sign of passing from *pre-contemplation* into preparation and action. This **future focus** is a central part of the Program. We spend some time working on futures that are accessible, motivating and self-validating. In order to do this, we attempt to build a sense of personal identity that is associated with a calling or life-direction. This idea, closely related to C. G. Jung's idea of *individuation* and Abraham Maslow's *self-actualization*, presumes that **every individual has a call or direction, an unconscious image of a personal developmental goal that can make life simultaneously challenging, exciting and <u>positively addictive</u>**. It is in the flow of this living stream that addictions are relativized into misadventures.

Self-esteem has been for many drug treatment professionals the elusive universal remedy for all manner of problems. Here, self-esteem flows out of a developing a sense of self-efficacy and the more intimate sense of self that evolves from the realization of the Call, or personal direction. In the past, we have misunderstood self-esteem as flowing from power, position, possessions or other more or less concrete entities. Here we understand that *Self-Esteem flows from a true knowledge of the Self.* It is in finding the essential core of Self and living harmoniously with that core calling that the Self receives the esteem it so vitally requires.

In the following pages we will guide you through a Program of self-actualization that provides specific tools for the development of self-efficacy, emotional control/maturity, self-discovery and future orientation. As a treatment Program provided to offenders with substance abuse issues in a criminal justice context, the drug treatment element was for the most part unspoken. However, it must be understood that every person in the group was sent there expressly *for* drug treatment. This approach had several benefits. First, it eliminated much of the resistance that is found in more traditional confrontational Programs. The participants all knew that the Program was about drug treatment but the issue was never made explicit beyond the first session. Second, because the exercises are empowering, enjoyable and inherently self-reinforcing, they provided an internally validated reference that tells the individual that *they are changing*, that *they have choices* and that *they* have value. All of the preachments, lectures and pep rallies in the world will never be able to compete with the sense of self-validating empowerment that comes from the direct, intuitive reality of personal experience. Third, rather than being forced or coerced into exercises aimed directly at the addictive problem and responding with resistance, or their own lack of treatment readiness, the exercises were shown to be of value to them as people. In the process of participating, each participant gained first-hand experience of their own flexibility and capacity to control their inner environment. As a result, *just as addictive behaviors generalize into non-addictive contexts, so our behaviors must generalize into the addictive circumstances.* Finally, among the less obvious advantages of the Program is its provision of observable success criteria. Besides keeping track of urine specimens (all participants were monitored), each exercise provides specific behavioral criteria for evaluating progress through the Program.

Applications of the Program.

The Brooklyn Program began life as a novel approach to substance abuse services. It is, however, important to realize that all of its roots and presuppositions are related not so much to drugs as they are to the principles of human growth as understood by Carl Jung, Abraham Maslow and the founders of Neuro-Linguistic Programming. The Brooklyn Program aims to enhance and restore optimal human functioning. It works for drugs because addictions are part of the normal range of human behavior as it manifests in abnormal contexts. The entire Program consists in the teaching and exploration of a set of cognitive and spiritual skills that are of universal relevance. As a result, the Program has applications that range far beyond substance use disorders.

Insofar as it begins with certain assumptions about human growth and development it should first be seen as a set of tools for personal growth that need no pathological context in which to function. One of the consequences of operating the Program within the precincts of the US Courts has been the need to treat fairly large numbers of persons who had no specific pathology other than finding themselves in the hands of the Courts for illegal behaviors. In such cases the diagnosis for treatment was typically made by uninformed but well-meaning members of the judiciary. For these persons, the Brooklyn Program provided a very specific context for personal growth in which the de-emphasis of drugs and substance abuse (after the first session) was a welcome reprieve from the unhappy possibility of weeks of unwanted and unneeded substance abuse education. For these participants, as for those with substance use problems, the main result of their participation was the awakening of a deep sense of Self, optimization of choice and a new sense of personal direction and worth. For most participants, the course becomes a radical awakening to the very best of what it means to be human.

The spiritual dimension is, on one hand, one of the more subtle attributes of the Program; on the other, it is the most dramatic. The Program is not overtly spiritual. It promises neither salvation nor enlightenment. It does, however, center the participant in the kind of place that is the sine qua non of advanced spiritual discipline. This link is easily followed through the works of Jung and Maslow and reveals itself as affective states are enhanced to archetypal intensity so as to instantiate what Maslow called "peak experiences". The end result is the constellation of a deep sense of Self that helps to generate new capacities to choose and the ability to move through the world more consciously, in a proactive manner. There are major affinities with classical Buddhism, the Gurdjieff work and other spiritual enterprises. Most importantly, the Program brings the participant to a place where they have a real sense of their own place in the universe. This is awakened first through the constellation of a deep sense of Self; it is later enhanced by integrating past experiences and future outcomes into the individual experience of a call or life direction.

Because the Program has no hidden agenda for conversion to any religious perspective, and because it provides this depth of experience through the medium of simple exercises in Western Psychology, the Program has been experienced as enhancing and supporting almost every kind of spiritual endeavor. Christians, Jews and Muslims report a greater centeredness in prayer and a deeper sense of the importan of their traditions. Taoists and Buddhists have found it awakening their own experiences of meditation and centering. For persons committed to a 12-step discipline, the deep sense of Self created in the Program becomes an effective operationalization of a higher power; especially where none other exists. For persons without a specific tradition, it provides a quiet centered place from which to reflect objectively on life. In the most secular sense it is the instantiation of Gendlin's Focusing. On a broader plain it is fully compatible with D'Aquili and Newberg's aesthetic – spiritual continuum.

In regard to pathology, emotional control and personal stability are natural fruits of the Program. Because we provide tools for actively choosing *how to feel* in order to meet the needs of various situations, the Program provides proven results for anger management. As one might guess, in a criminal justice population, there are ample numbers of violent and angry offenders. One of the more consistent reports that we have received from participants and their probation officers is that they have calmed down and are making better choices.

Attention deficits are often ameliorated in the process of completing the exercises. By using techniques similar to mindfulness meditation that result in focused, highly pleasurable states of mind and body, participants learn to focus their attention inwardly. With the use of self-anchors or conditioned stimuli, this capacity to focus can be transferred to any context.

There are no panaceas. The Brooklyn Program is not a cure-all. It does, however, provide psychological tools that are consistent with optimal patterns of human growth and development. In those cases where psychological problems are without significant organic base, the Brooklyn Program can often awaken resources from the participant's behavioral repertoire to meet the present need. For persons already embarked in a productive life, it provides clear guidance for the path ahead and tools for assistance along the way.

Organization of the Text.

This version of the Manual has been reorganized in order to make the exercises more immediately available to the reader. A large amount of material aimed at specific replications of the Program in a governmental or drug treatment setting has been eliminated in order to provide more direct access to the exercises themselves. The exercises have also been modified to make them easier to use outside of the Programmatic context. This also has the effect of making the materials more available as a general Program for personal growth and development. In doing so, it is the author's hope that people using the manual will try some of the techniques; even if the Program as a whole does not meet their needs. A short bibliography and practical applications have been added at the end of each exercise. These are intended to provide further information for interested participants and to stimulate creative uses of the materials. Complete references appear at the end of the manual.

A conceptual introduction (Thinking About Drugs and Addiction) has been added to provide a theoretical base for the materials that follow. The original version of this chapter was published online by NLP Comprehensive. It remains the best introduction to the underlying neurological principles upon which the Program depends. Parts One and Two review the specific exercises and the techniques used in the Program. Part Three reviews the meditations and scripts for the exercises.. Each section begins with an overview and introduction. The chapters within each section present the exercises, one at a time, with descriptions of the presuppositions upon which they are based, instructional guidelines, success criteria and the text of the exercise. The same procedure is followed for the meditations. Part Four, Application Notes, provides further, point-by-point guidance through the first nine sessions. These notes provide significant, in-depth guidance for operating the Program through a full 16-week course and add instructions beyond those provided in the basic text. I have added a typical timeline for progress through the program. I have also made additions to the One-on-one Session based upon some kind suggestions by Steve Andreas. An extensive reference section appears at the end.

Richard M. Gray, Ph.D.

Highlands, NJ

June 8, 2011

Acknowledgments

The following volume owes a debt of gratitude to many people. First I would like to thank my wife, Florence Tomasulo Gray, whose continuing encouragement and faith made all of this possible. Without her steadfast belief in me and what I do, I could never have placed the long hours of work and study that brought the Program to fruition. Thank you, My Love.

Next, my thanks go to Jim Fox who was Chief Probation Officer in Brooklyn as the Program was developed. He provided the crucial permissions and support that made this possible and allowed the Program to operate under the auspices of the US Probation Department.

At the same time, I cannot forget my good friend and immediate superior throughout the development of the Program, George Doerrbecker. During most of my tenure with the Federal Government, George has been a friend, an advocate and a faithful evaluator of all of my ideas. When things got rough, George made a way through. I am confident that without his advocacy, this could not have happened.

At the time the Program began, and throughout its operation, I enjoyed the friendship and support of SUSPO Larry Cavagnetto. Larry worked with me in presenting the Program to offenders and worked hard to insure that it was a success. He has used the Program with many offenders and has been a good friend and a great support.

Further back in time, Steve Rackmill and Ralph Kistner provided me with the opportunity to obtain the training that made all of this possible. They gave me time to learn NLP as part of my graduate studies. For that I am grateful.

More directly, I would like to thank Anne' Linden for her fine training in NLP. I also need to acknowledge my good friend A. Stanley Cunningham. For several years after completing our training in NLP, Stanley and I met weekly to practice and work out new patterns. The root patterns in this Program and several of the exercises were originally developed with his help. He is a dear friend and a trusted partner.

I would also like to acknowledge the kindness of Leslie Lebeau and David Gordon for honoring me with the NLP World Community Award in Education. Much more than the award, their kindness and support has been a great blessing. I am thankful every day for their confidence.

I would like to thank Steve Andreas for taking the time to read through the manual and provide some helpful suggestions. I have come to cherish his friendship.

I would also like to thank Robert Dilts for giving me permission to use Sponsoring a Potential (Exercise Twelve).

Without the contributions of Leslie, David, Steve and Robert, NLP would not be quite what it is today.

I would like to also thank Win Wenger for permission to use his ImageStreaming. Although we have never met in person, I have been inspired by his enthusiasm and generosity.

Special thanks go to Steve Leeds and Rachel Hott of the NLP Center of NY. I have taken many great trainings there, more importantly, however, their recommendation for the NLP World Community Award was a crucial element in all of what has been happening in the last several years.

Similarly, my thanks go to Tom Dotz who, despite having never met me, recommended my work for the award.

Finally, my thanks go to Richard Bandler and John Grinder. I have not met either, but their work lies at the heart of anything having to do with NLP. My thanks to them for making all of this possible.

Introduction: Thinking About Drugs and Addiction[1]

One of the central difficulties in thinking effectively about drugs and addictions lies in the fact that the whole business is befuddled by propaganda, outdated information and superstition. NLP brings some wonderful tools to the discussion but before those tools can be effective, we need to clear the air.

Whenever people begin to talk about drugs (including alcohol) the following ideas are trumpeted as fact:

- Addiction is a progressive chronic disease that ends in death.
- There is some identifiable thing called addiction.
- All drug use inevitably leads to addiction.
- Certain drugs have the specific property of being addictive.
- Certain people are born with addictive personalities.

---

[1] This chapter is adapted from an article published in 2005 by NLP Comprehensive at: http://www.nlpcomprehensive.com/articles/AddictionsGray.html.

In a real sense, not one of these ideas is 'true'. Each of them is a generalization or distortion that proceeds from the medical and moral models of addiction that have framed most of our thinking about drugs and alcohol. They have also crippled our capacity to deal effectively with the concept of addictions.

Anciently, and far into the Twentieth Century, addiction was treated as a moral failing. It was a sin or error of excess. It was proof positive of the presence of some lack of personal virtue, self-control or will power. Addicts were sinners or idiots. Through most of American history, the addict's ruin was viewed as just desserts but the toll taken on long-suffering family members and business associates was scandalous. (Shattuck, 1994).

In the 1930s, when Bill Wilson and his associates put together the basic ideas of Alcoholics Anonymous, they decided that alcoholism (and later, drug addiction more generally) should be treated as a disease of the spirit. They held that alcoholism, while rooted in moral failings and character defects, had its final manifestation as disease.

Although the disease concept was originally designed as a metaphor with the intent of saving addicts from humiliation, in 1956 the AMA accorded alcoholism the medical status of disease. From then on the idea of addiction-as-disease gained momentum and was finally concretized through the growth of the huge business concerns that developed around it. The medicalization of addiction came to full fruition in the Minnesota Model which immortalized the definition of addiction as a chronic, progressive disease that ultimately ends in death (Laundergan, 1983; Doweiko, 1996; Mann & Heinz, 2000; Peele, 1989; Peele & Brodsky, 1991).

There are multiple reasons for arguing against the idea that addictions are diseases. Here, however, we will focus only on the fact that the disease concept implies a level of brokenness and biological stasis that limits creative thinking about the problem. From an NLP perspective, and from the perspective of a growing body of neuroscience, it may be useful to think of addictions as a set of over-learned and over-valued behaviors.

In a later section of the chapter we will look at how addiction works and how it affects values hierarchies and preference criteria. When we look at addictions from that perspective we will see that they are not diseases but powerful preferences.

The Thing We Call Addiction

Nominalizations are powerful distortions of reality. By turning a set of actions or symptoms into a static label we often miss the dynamic reality of the problem itself. In the case of addictions, we are often so blinded by the label that we miss the underlying utility of the behavior and the fact that it serves or has served some practical use in the life in question. In NLP these are described as positive intentions (O'Connor & Seymour, 1990; Bandler & Grinder, 1975).

Over the course of a lifetime, 'addictions' ebb and flow. For some persons, allegedly chronic, progressive addictions disappear for years and then suddenly reappear. For others, a terrible addiction suddenly goes away forever. These are not the behaviors of *things*. They are qualities of concepts; expressions of personal behaviors that are active in some contexts and dormant in others. Like other behavioral preferences, addictions are bound to contexts. Contexts may relate to persons, places, things, environments, self-definitions and mood.

In many cases, just being away from certain people or contexts ends the need to engage in the behavior. For others, the geographical cure, moving away, works. Stanton Peele reports how the Veterans Administration prepared huge resources to meet the anticipated flood of addicted GIs returning from Viet Nam. They knew that many of our soldiers had developed significant heroin addictions while in the service and expected that they would need a great deal of help when they returned home to the US. When they returned, however, levels of addiction dropped dramatically. The original report of the study indicates that an addiction of 43% in Viet Nam dropped to 4% when the GIs returned stateside. Diseases do not respond to social context (Gray, 2008; Peele & Brodsky, 1991; Robins, Davis, & Nurco, 1974).

For our purposes, it may be useful to think about 'addictions' as reifications of patterns of behavior. They are not real things, but things that are solidified into illusory realities by the words we use. They are conjured into existence by the labels we apply to them. What happens when we begin to think of preferences and skills instead of diseases?

If we think of an addiction as a set of (often unconscious) preferences, we may then be able to discover another set of preferences that are more valuable than the problem substance or behavior. If we know the utility of the problem behavior, it may be possible to find behaviors that are more useful, more immediate and more intuitive than these others. This analysis was suggested by Bandler and Grinder in the mid-1970s. If we can find an ultimate set of criteria, we stand the chance of 'outframing' the entire problem (Bandler & Grinder, 1979, 1982; Hall, 1996).

If we begin to think of addictions as learned skill sets that come to be preferred patterns of action, what happens if we find a generalized pattern of behavior that works more effectively? If we create a new set of behaviors, beliefs or experiences that serve the purposes of social integration, positive self-regard, transcendence, etc., what happens to the addiction?

One of the interesting facets of the interface between the concepts of addiction-as-disease and addiction-as-skill-set is that you never really lose a skill. Once it's learned, it can always be revived. It may take a little practice but, like bicycle riding, it is always there. How does this differ from the chronic nature of addictions? If addiction is the skill of solving every problem by artificially, and momentarily transcending it, is it unreasonable that when other strategies fail, the problem behavior recurs? Just so, if the new behaviors are sufficiently rewarding, the old skill may never be needed again.

The idea that drug use inevitably leads to addiction is conceptually the same as saying that kissing makes you pregnant. While it is necessarily true that you have to have the first drink in order to become an alcoholic, it does not follow that having a drink will inevitably lead you to alcoholism. Modern neuroscience has shown clearly that addicts do not necessarily pass through the stages of use, abuse, dependence and addiction.

Part of the problem here lies in the demonization of mind altering substances. Because our culture lacks appropriate rules and guidelines for the use of psychotropic substances, we project upon them the shadow of the unknown. Because our culture focuses so insistently on conscious process, every appeal to unconscious process is viewed with suspicion (Furst, 1976; Gray, 1996, Zoja, 1996).

Addiction is not a property of chemical agents, it is about how people use the substances and behaviors. One of the key things that modern neuroscience tells us is that addiction is a property of brains that are functioning normally.

Neuroscience and addiction

In recent years, cognitive neuroscience has shed significant light on the problems of addiction and substance abuse. These researches have uncovered a close relationship between drug addictions, behavioral addictions, compulsions and more normal patterns of reward and motivation. Central to this information are the ideas that drug and behavioral addictions are not problems with the 'hedonic impact' of the reinforcing agent ('liking' the drug), but they are problems related to 'wanting' or 'craving' the agent. They have called the measure of wanting, incentive salience. A second important discovery is that the mechanism of craving or incentive salience is mediated by neurons in the midbrain that produce dopamine. The midbrain dopamine tract runs from the Ventral Tegmental Area at the base of the brain; through the Nucleus Accumbens, at the base of the Ventral Striatum; and finally ends at the Orbito-Frontal Cortex, the apparent control center for motivation and wanting in general (Robinson, 2004; Robinson & Berridge, 2001; Ruden, 1997; Schultz, Dayan & Montague, 1997; Tobler, Fiorillo & Schultz, 2005; Waelti, Dickenson, & Schultz, 2001)

For a long time, it was believed that people got hooked on drugs or other substances and behaviors because they felt good. While this is certainly part of the reason, it doesn't explain why, even after substances or behaviors cease to produce the same 'whack', people continue to seek them out. The 'feeling good' interpretation of behavioral addiction violates some of the cardinal principles of behavioral psychology. Every addictive drug, every behavioral addiction, and every learned behavior is subject to habituation. This means that the more exposure you have to something, the less effective it becomes. When a behavior ceases to be rewarding, the behavior becomes less probable. At some point, the stimulus stops evoking the trained response and the response is said to have been extinguished.

By this rule, most substances of abuse and most behavioral addictions should disappear on their own as they become less and less rewarding. However, even though, over time, addicts report lessened pleasure from the drug or behavior (decreased hedonic impact), they complain that they still want the drug. This has led researchers to focus not on the pleasure that drugs impart (hedonic impact) but on their ability to create craving or wanting (incentive salience). It is this factor, craving or wanting, that is mediated by the midbrain dopamine system (O'Brien & Gardiner, 2005; Robinson & Berridge, 2001).

Incentive salience connects to neurophysiology through a series of experiments on single dopaminergic neurons and neural implants measuring the response of the neurons to various stimulus conditions. In general, researchers found that the midbrain dopamine system responds in very specific and predictable ways. First, it responds powerfully to novel rewards. Whenever rewards appear in an unexpected context, these neurons respond vigorously. Second, the brain seeks "the difference that makes a difference". If a stimulus fully predicts a reward or if it predicts decreasing reward, the neuronal response decreases and often disappears (This is the neural root of habituation.). Third, if the stimulus predicts a reward that appears reliably but increases in value relative to other recent rewards, the neurons again increase the intensity of their response (Robinson, 2004; Robinson & Berridge, 2001; Schultz, Dayan & Montague, 1997; Tobler, Fiorillo & Schultz, 2005; Waelti, Dickenson, & Schultz, 2001).

This data relates to addiction in the following ways:

Novelty is a crucial part of the value accorded to addictive behaviors and substances (Robinson, 2004; Robinson & Berridge, 2001; Schultz, Dayan & Montague, 1997; Tobler, Fiorillo & Schultz, 2005; Waelti, Dickenson, & Schultz, 2001). From a Jungian perspective, Luigi Zoja (1990) links addiction to a failed attempt at transcendence. Addicts and drug abusers are often seeking a new spiritual perspective but end up trapped in a consumerist nightmare. There is a huge literature on relapse prevention pointing to boredom and stress as crucial predictors of relapse. In standard behavioral literature, animals who have suffered sensory deprivation will perform for rewards consisting of nothing more than exposure to novelty (Chambers, Bickel, & Potenza, 2007; Daly, Mercer & Carpenter, 2002; Franken, 2003; NIDA, 2002).

Inconsistency of reward reflects the standard behavioral idea of schedules of reinforcement. Once behaviors have been established through simple reinforcement, their probable repetition can be enhanced by changing the frequency or schedule of reinforcement. That is, instead of rewarding every correct trial one might reward every third correct response or every response that happens 10 seconds after the first. This kind of reinforcement schedule is associated with persistent and sometimes compulsive behaviors (Ferster & Skinner, 1953; Skinner, 1957). In the world of addiction, the initial encounters with the drug providers do not always provide access to the high. The contexts of the problem behaviors do not always reliably predict access (the wrong company, the wrong place, nothing available). Drug cues in general set up an expectancy that is not always fulfilled. This very inconsistency increases the power of expectation.

The relative intensity of addictive behaviors, as compared to normal experiences, leads the substance abuser to anticipate and prefer them over more mundane rewards. One of the important things about this observation is that the comparisons made by the dopamine systems are short term. Behavioral preferences are established when there is a sharp difference in intensity between problem reinforcers and other recently experienced stimuli. Positive experiences from last month are not remembered in the context of a drug or behavior that that is overwhelmingly better than anything in the last hour (Chambers et al., 2007; Schultz, Dayan & Montague, 1997; Tobler, Fiorillo & Schultz, 2005; Waelti, Dickenson, & Schultz, 2001).

Addictive behaviors tend to appear after intense exposure to the substance or behavior. Although there may be such things as one-shot learnings of addictive responses (even though this is highly unlikely), the compulsive behaviors called addictions tend to be established over multiple experiences, especially when those experiences are repeated with great frequency over a short time. Addictions are rarely established in less than a year of intense use (Inaba & Cohen, 2007; Robinson, 2004).

Another important insight from neuroscience should be familiar to practitioners of NLP. Preferences and values are experienced hierarchically. That means that we accord more or less value (incentive salience) to various actions and experiences. In NLP we describe these preference hierarchies in terms of value criteria. According to most researchers, the problem of addiction consists most centrally in the fact that the addictive behavior or substance is so far over-valued that it 'outframes' normal response systems (Berridge & Robinson, 2003; Dilts & DeLozier, 2000; Goldstein & Volkow, 2002; Kringleback, 2005; Kringlebach & Berridge, 2009; McClure, Daw & Montague, 2003; O'Connor & Seymour, 1990).

As noted previously, the midbrain dopamine system responds to the most impactful stimulus in recent neural history. Drugs, risky behavior, shoplifting, chocolate and sex often provide a significantly more powerful experience than many other behaviors that we encounter daily. As a result, they are promoted to the top of the preference hierarchy (Chambers et al., 2007). This promotion happens in two ways:

With addictive substances, the primary means by which using behaviors are accorded increased incentive salience is through the direct or indirect chemical action on the midbrain dopamine system. Whether directly (like amphetamines) or indirectly (like alcohol or heroin), substances of abuse create an inordinate output of dopamine that tells the brain, "This is really important!" and "We need to do this much more often!"

The second way that behaviors are promoted in the hierarchy, is through behavioral adaptations. The same midbrain dopamine system is activated whenever a particular outcome or behavior can be used: 1. as an integral part of different behavioral sequences ("I always have a drink before I go out, just to loosen up." "Whenever I have to face John's mother, I have a drink."). In the language of behavioral science we would say that the behavior is present in multiple schemas. 2. It is found to be useful or available in multiple contexts (Cigarettes and alcohol become powerfully addictive because they are so well integrated into the contexts of everyday life.). 3. A behavior becomes important when it seems to represent an easy answer, the path of least resistance. Drugs and behavioral problems work quickly and effectively to remove the stressors of the moment. They are easy, if impermanent, answers (Austin & Vancouver, 1996). In effect, the short term utility of the behavior and its generalization into multiple contexts tells the brain, "This is important!"

Recent work in Neuroscience has indicated that substance abusers also suffer from inhibitory deficits related to malfunctions in the Frontal Cortex, where target behaviors are identified and attention is focused and in the Anterior Cingulate Cortex which monitors performance and risk. Some of those same studies suggest that reactivation of the frontal cortex can ameliorate some of the problems associated with addictive cravings (Bechara, 2005; Bechara & Damasio, 2002; Bechara, Damasio, & Damasio, 2000; Bechara, Damasio, Damasio, & Lee, 1999; Bechara, Dolan, & Hindes, 2002; Bechara, & van der Kooy, 1985; Feil, Sheppard, Fitzgerald, Yücelc, Lubman, & Bradshaw, 2010).

So far, we have identified two means by which behaviors and substances can be promoted to the top of the preference hierarchy so that they outframe other behaviors and preferences. How is this important for an NLP based understanding of addiction?

It is important precisely because what we observe in an addictive behavior is an expression of a fairly normal value hierarchy created under extraordinary circumstances. It tells us not that the brain is broken, but that it is doing what it always does: prioritize behaviors in terms of their immediate utility for the organism. If this is so, then the preference hierarchy is not only the locus of the addiction problem, it is also the key to eliminating the problem. This also helps us to understand why certain things like spiritual experience, falling in love, finding a personal direction and even just "getting out of Dodge" can work. They work because they represent criteria or experiences that are more highly valued than the experience of the addictive substance or behavior. They outframe the problem.

NLP Approaches to Addiction

In the early history of NLP, Richard Bandler and John Grinder made several suggestions about the treatment of addiction. In Reframing (1972), they suggest that addictions could be treated by providing the client with a response option that was more powerful, more accessible and more immediate than the drug itself. This statement was one of the root inspirations for the Brooklyn Program. In several sources, Bandler suggests making the state of being high available as an anchor or intensifying the urge to use to the point where it becomes preferable to the use itself (Bandler, 1997, 1999).

Steve Andreas suggests using the compulsion blow-out to solve the immediate problem of craving. He also suggests using the guilt resolution process and other techniques used for clean up of motivations and secondary gain (Andreas & Andreas, 1979, 2002).

One of the early applications of NLP to the treatment of addictions was the six step reframe (Bandler & Grinder 1979, 1982). This technique was promulgated specifically for use in addictions by Shelly Sternman in her 1990 book, Neuro Linguistic Programming in Alcoholism Treatment.

The six step reframe, as one of the old standbys, continues to be used for the treatment of addictions and works by enlisting the aid of the unconscious mind, as personified in the 'part' responsible for the presenting problem, to generate more useful alternatives that will realize the original positive intent of the behavior. The process has been critiqued in terms of its tendency to artificially fragment the personality and its general allopathic orientation.

Another, more elegant approach to addictions was provided by Connirae and Tamara Andreas in their 1994 book, Core Transformations. This approach looked to uncover a series of outcome sequiturs from the problem behavior that would eventually lead to deep, core-level values and experiences. These core values could be understood as the ultimate positive intent of the behavior. Once conscious, the core value could become the active outcome towards which organismic energies would be directed.

From a Jungian and generative perspective, both of these approaches reach down to access an archetypal level of experience that can be used to redirect conscious and unconscious energies in a direction that is much more aligned with archetype of the deep Self—the center and goal towards which each life tends to grow (Gray, 1996, 1997). In the six step reframe, the approach is accomplished outside of consciousness with the expectation that all of the parts (presumably Jungian complexes -- behavioral and perceptual habit centers) will be able to negotiate an effective, alternative answer to the problem behavior and then replace it. The approach of Core Transformation works from the problem behavior to reach one of several possible conscious experiences of wholeness, a core value that serves to provide a new direction for the behavior. In this approach, something much closer to a conscious experience of the Deep Self is awakened and acknowledged.

Both interventions, however, because of their starting point in pathology, are stuck with the relatively incomplete instantiation of unity and several layers of objections from parts that must be dealt with in order to finalize the process. In light of this, the Brooklyn Program, which drew inspiration from Core Transformations, asked: "How can we uncover the unitary Self, the deep personal direction that Core Transformation successfully uncovers and how can we make it the defining context of behavior without beginning with the problem state?"

Eventually, the Program was designed so as to structure a complex anchor that would awaken (or 'constellate,' in Jungian terms) a sense of the deep Self consistent with the ideas of personal growth set forth by C. G. Jung and Abraham Maslow. This resource would serve to provide a state that would not only outframe the addictive process but would also center the individual in a life that he or she would find meaningful in a continuing manner over time (Gray, 2001, 2002, 2003, 2005; Hillman, 1996; Jung, 1979, 1984; Maslow, 1970).

Such a state could be created by bringing together a series of positive life experiences. By anchoring the felt sense of those experiences, and stacking those anchors together into a single (anchored) resource, there would emerge a single, positive affect representing the deepest and most positive aspects of the individual. If the exemplars for the anchors were correctly chosen, they would provide a sense of growth into the center of personal potential that would serve to awaken a meaningful life direction. Because the Jungian dynamic of archetypes and complexes (the intellectual source of this structure) exists at the level of inchoate, felt experiences, an anchoring procedure was deemed to be the perfect means for creating the state and making it available (Gray, 1997, 2001, 2002, 2003, 2005; Jung, 1979, 1984).

In 1997, while still working with a program of anchored resources in the context of a rather standard cognitive program, the author had the good fortune to attend a workshop on the midbrain dopamine system. The net effect of this presentation was to outline some of the early research already noted above. More importantly, the view from physiology made me realize that substance abuse treatment could be handled completely and effectively on the basis of feelings alone. One did not have to convince anyone of anything, all of the change could be produced by providing a series of positive, ecstatic experiences that were more accessible, more intuitive and more valuable to the client; neuroscience seemed to be confirming NLP. If the states also provided more flexibility and opened future options, their utility would be enhanced.

The program finally came to take on the following characteristics: After a brief introduction to the nature of addiction and the hierarchy of salience; drugs, addiction and problem behaviors were never formally mentioned again during the entire 16-week program. The early emphasis of the program turned from the central archetypal theme of awakening the deep Self, to teaching the participants how to create a series of powerful ecstatic states over which they exercised total control. A continuing part of the emphasis called for those states to be anchored in an easy and repeatable way so that each of them could be elicited at will, and its intensity manipulated. The outcome of increased flexibility was realized by suggesting that participants experiment with the anchors in multiple contexts. In this way they could experience for themselves the utility of the anchors and their independence from the facilitator and the treatment context (Gray, 2001, 2002, 2003, 2005).

In general, the program sought to teach the art of ecstasy and its use for transcendence. It was continually presented as a refuge from the cares of the world; two hours a week where you could "always leave feeling better than you did when you came in". It was a place where no one was preached at, demeaned or questioned; and it all happened at the Probation Department.

Once a root set of anchors were taught (The list was borrowed from Carmine Baffa's (1997) web site), participants were encouraged to use them in their everyday lives to relax, feel better, and to gain control. Participants found that while using the anchors, negative behaviors like road rage and impatience tended to disappear. They often never consciously realized that their drug and alcohol-related urges had disappeared until well into the program. It was often not until the last sessions that participants realized not only that we had not discussed drugs, but that they had experienced few if any cravings.

After the root skill of anchoring ecstatic states was completed, participants were urged to create a deep sense of Self by stacking those anchors.

In the context of creating states designed to establish, or re-establish a values hierarchy in order to 'outframe' addictive behaviors, there are two technical insights that proceed directly from experience in the

Brooklyn Program and accord well with cognitive neuroscience. The first is this: If we structure a positive experience or experiences, so that it will compete successfully against a problem state, the competing experiences must be valued for their own sake, not in their instrumental relationship to the problem behaviors. The second is: The competing behavior must point to or promise a positively motivating future.

The first of these insights comes again from work related to the midbrain dopamine system and the frontal lobes. We have all been exposed to the functions (and supposed functions) of the left and right hemispheres of the brain. For our purposes, one of the more important aspects of hemispheric lateralization points to the idea that positive (or approach-valenced) experiences are processed in the left frontal lobes and negative (or withdrawal-valenced) experiences are processed in the right frontal lobes. This means that positive choices, the evaluation of what we want, the hierarchy of preferences, is processed in a totally different place than the measures of what we don't want. It is a crucial piece of information (Craig, 2009; Davidson, 1993; Davidson & Harrington, 2002).

Similarly, LeDoux (2002) points out that the brain has many circuits, many run in parallel while others are mutually exclusive. Positive and negative affect are mutually exclusive (Ambivalence is alternation between the two and is handled by another part of the brain, the Anterior Cingulate Cortex.). In order for an anchor or other positive affect to have maximum utility, it must be developed and used as a positive good in itself, not as a tool for dealing with a problem. When we use it as a tool for a specific purpose, create it as a tool to combat addiction, or otherwise associate it with a negative outcome, we lose some of its utility. For this reason, we emphasize that all of the exercises in the Brooklyn Program must first be pursued for their own sake and, insofar as possible, with no reference to controlling—or stopping—anything (Gray, 2008).

The second insight comes both from neuroscience and from the work of Prochaska, Di Clemente and Norcross (1994) in their Stages of Change Model. The Stages of Change Model is probably the most frequently used scientific model for personal change in the world today. James Prochaska, the lead author, made a crucial insight that moots much of the process that he discovered. He indicated that all of the change from what he calls precontemplation (classic denial in old-style parlance) to action (doing something about the problem) correlates--in every kind of change--to one thing: *wanting something that is more important than the problem behavior.* He further noted that disliking the problem behavior does not cause the change. The devaluation, or disliking, of the problem behavior arises as a response to wanting or liking something better or more important. Devaluing the problem behavior, if it occurs, is a result of effective change work, not its source.

Prochaska's insight points us back to value hierarchies, and once again to the awakening of positive affect as a means of re-sorting that hierarchy. It integrates with the neurophysiology discussed above as follows: 1. Wanting something passionately has the capacity to reset the preference hierarchy. 2. Wanting something for its own sake, devalues the unwanted behavior as a consequence of that resorting. 3. Because the mechanisms of the frontal lobes for wanting and avoiding are separate, it is most important to build intense, positive states and motivators in order to propel the change. 4. Because positive states point to positive futures, they can and should be structured as a path of development; a set of developing outcomes. This will have the effect of frustrating the brain's tendency to habituate and will keep the dopamine system registering that goal as more highly valued. 5. Because such positive experiences reawaken the left Orbito Frontal Cortex to preferences that are independent of drugs or destructive behavior, they hold forth the hope of restoring frontal function that substance use has inactivated.

These insights are supported in the work of Reddish, as reported by Chambers et al. (2007). These authors indicate that effective challenges to the neuro-cognitive networks that represent addictive behavior can only be successfully mounted by the creation of independent, self-maintaining behavioral networks that support other options. That is they must appeal to separate, more fundamental values that exist independent of the addictive frame.

A crucial technical refinement in the program regards the process of anchoring. In order to create states that are useful across contexts, participants are taught to anchor states that are, as far as possible, devoid of content. Anchors that retain contextual information have limited utility. If, however, in the process of creating and anchoring the state, contextual information is reduced so that all of the attention is placed on a disconnected, ecstatic, floating trance, carrying at most the felt tone of the original experience, the anchor can be used anywhere and that same anchor can be used to create fully integrated complex states rather than crude aggregates of unrelated experiences. This makes the actor and the action the context of use, not any circumstance or memory (Bandler, 1999; Gray, 2008).

The Brooklyn Program achieves this by overloading the capacity of short term memory using the principle of Miller's Magic Number (Miller, 1956). By increasing the amount of short-term attention invested in the felt sense of the experience (How does it feel? Where is the feeling? How does it move? What is the texture of the feeling? Does the feeling make a sound? What color is the feeling? What is its temperature? Where is the warmest part?), one can allow all of the contextual information in a remembered experience to fade away (Gray, 2008).

The other, and final refinement, is our focus on the crucial role of ecstasy. Positive feelings reawaken the left orbito-frontal cortex and with it, the capacity to choose. The reawakening of Frontal function has been suggested as a key to overcoming addiction. One of the most important skills taught in the Brooklyn Program is the possibility and the skill of feeling wonderful. Most of the time spent in the program is spent accessing positive states; finding ways to explore them, and making sure that they become expanding paths of discovery, rather than destinations (Feil, Sheppard, Fitzgerald, Yücelc, Lubman, & Bradshaw, 2010; Gray, 2005, 2008).

In general, it has been my aim to give people back to themselves. The Brooklyn Program, above all else, seeks to restore levels of choice, self-esteem (rooted in a deep sense of who I am and must be becoming) and joy that can serve as the root of personal growth into the future (Gray, 2001, 2002, 2003, 2005, 2005a).

REFERENCES

Andreas, C. & Andreas, S. (1989). *Heart of the mind.* Moab, UT: Real People Press.

Andreas, S. & Andreas, C. (1987). *Change your mind-- and keep the change.* Moab, UT: Real People Press.

Andreas, C., & Andreas, T. (1994). *Core transformations.* Moab, UT: Real People Press.

Austin, J. T. & Vancouver, J. B. (1996). Goal constructs in psychology: Structure, process, and content. *Psychological Bulletin, 120*(3), 338-375.

Baffa, C. (1997). *IQ, Hypnosis and Genius.* HTTP://Carmine.net/geni/geni0001.htm

Bandler, R. & Grinder, J. (1975). *the structure of magic I.* Cupertino, Calif.: Science and Behavior Books.

Bandler, R & Grinder, J. (1979). *Frogs into princes.* Moab, UT: Real People Press.

Bandler, R. & Grinder, J. (1982). *Reframing: Neuro-Linguistic Programming and the transformation of meaning.* Moab, UT: Real People Press.

Bandler, R. & MacDonald, W. (1987). *An insider's guide to submodalities.* Moab, UT: Real People Press.

Bandler, R. (1985). *Using your brain for a change.* Moab, UT: Real People Press.

Bandler, R. (1993). *Time for a change.* Capitola, CA: Meta Publications.

Bandler, R. (1999). *Introduction to DHE.* Chicago (Audio).

Bandler, R. (ND). *The genius of Richard Bandler.* Boulder, CO: NLP Comprehensive (Audio).

Bechara, A.; Damasio, H., & Damasio, A.R. (2000). Emotion, Decision Making and the Orbitofrontal Cortex. Cerebral Cortex, vol. 10, No. 3, 295-307

Bechara A., & Damasio, H. (2002). Decision-making and addiction (part I): Impaired activation of somatic states in substance dependent individuals when pondering decisions with negative future consequences. *Neuropsychologia. 40*(10), 1675-1689.

Bechara, A., Dolan, S., & Hindes, A. (2002). Decision-making and addiction (part II): Myopia for the future or hypersensitivity to reward? *Neuropsychologia. 40*(10), 1690-1705.

Bechara, A., Damasio H., Damasio, A., & Lee, G. (1999). Different contributions of the human amygdala and ventromedial prefrontal cortex. *The Journal of Neuroscience, 19*(13), 5473-5481.

Bechara, A. (2005). Decision making, impulse control and loss of willpower to resist drugs: a neurocognitive perspective. *Nature Neuroscience, 8*(11), pp 1458 – 1463.

Bechara, A., & van der Kooy, D. (1985). Opposite motivational effects of endogenous opioids in brain and periphery. *Nature, 314*, 533-534.

Berridge, K. C., & Robinson, T. E. (2003). Parsing reward. *Trends in Neuroscience. 26*(9), 507-513.

Chambers, R. A., Bickel, W. K., & Potenza, M. N. (2007). A scale-free systems theory of motivation and addiction. *Neuroscience & Biobehavioral Reviews , 31*(7): 1017-1045.

Craig, A. D. (2009). How do you feel--now? The anterior insula and human awareness. *Nature Reviews Neuroscience 10*(1): 59-70.

Daley, D. C., Mercer, D., & Carpenter, G. (2002). *The collaborative cocaine treatment study model.* Bethesda, MD: NIDA.

Davidson, R. J. (1993). Parsing affective space: Perspectives from neuropsychology and psychophysiology. *Neuropsychology, 7*(4), 464-475.

Davidson, R. J., & Harrington, A. (2002). *Visions of compassion: Western scientists and tibetan buddhists examine human nature.* NY: Oxford University Press.

Dilts, R., & Delozier, J. (2000). *the encyclopedia of systemic neuro-linguistic programming and nlp new coding.* Scotts Valley, CA: NLP University Press. Retrieved at www.nlpu.com

Doweiko, H. (1996). *Concepts of chemical dependency (Third Ed.).* Pacific Grove, CA: Brooks/Cole.

Feil, J., Sheppard, D., Fitzgerald, P. B., Yücelc M., Lubman, D. I., & Bradshaw, J. L. (2010). Addiction, compulsive drug seeking, and the role of frontostriatal mechanisms in regulating inhibitory control. *Neuroscience and Biobehavioral Reviews, 35,* 248–275.

Ferster, C. B. & Skinner, B. F. (1953). *Schedules of reinforcement.* NY: Macmillan Free Press.

Franken, I. H. A. (2003). Drug craving and addiction: integrating psychological and neuropsychopharmacological approaches. *Progress in Neuro-Psychopharmacology and Biological Psychiatry, 27*(4), 563-579. doi:10.1016/S0278-5846(03)00081-2.

Furst, P. T. (1976). *Hallucinogens and culture.* Novato, CA: Chandler and Sharp.

Goldstein, R. Z., & Volkow, N. D. (2002). Drug addiction and its underlying neurobiological basis: Neuroimaging evidence for the involvement of the frontal cortex. *American Journal of Psychiatry, 159*(10).

Gray, R. M. (1997a). Ericksonian approaches to the ego-self axis: Establishing futurity and a sense of self in addictive clients. *Innovative Approaches to the Treatment of Substance Abuse for the Twenty First Century.* St. Francis College, Brooklyn, NY. Published on the WWW at http://www.temperance.com/nlp-addict/articles.html

Gray, R. M. (1996). *Archetypal explorations.* London: Routledge.

Gray, R. M. (2001). Addictions and the self: A self-enhancement model for drug treatment in the criminal justice system. *The Journal of Social Work Practice in the Addictions, 2*(1).

Gray, R. M. (2002). The Brooklyn Program: Innovative approaches to substance abuse treatment. *Federal Probation Quarterly, 66*(3).

Gray, R. M. (2003). The Brooklyn Program: Cognitive applications of the physiological correlates of spiritual experience (Conference Workshop). Presented at the Dr. Lonnie E. Mitchell National HBCU Substance Abuse Conference, sponsored by Howard University, on April 2, 2003. http://richardmgray.home.comcast.net

Gray, R. M. (2004). *Incentive salience, meditation and the neurobiology of addiction* (Conference Workshop). Presented at The Dr. Lonnie E. Mitchell National HBCU Substance Abuse Conference. Baltimore, MD, April 1, 2004.

Gray, R. M. (2008). *Tranforming futures: The Brooklyn Program facilitators manual*. Raleigh, NC: Lulu Press.

Gray, R. M. (2005). Transcending context: Awakening the physiology of spirit Unpublished manuscript based on a paper presented at the National Association Of Social Workers Annual Addictions Institute, May 2, 2004.

Hall, L. M. (1996). *Meta states: A Domain of logical levels*. Grand Junction, CO.: Empowerment Technologies.

Hillman, J. (1996). *The Soul's code: In search of character and calling*. NY: Random House.

Inaba, D., & Cohen, W. E. (2007). *Uppers downers and all arounders (7th ed.)*. Medford Oregon: CNS Press.

Jung, C. G. (1979). *The archetypes of the collective unconscious. (CW9i)*. Princeton: Princeton Univ. Press.

Jung, C. G. (1984). Psychology and western religion (CW11). Princeton: Princeton Univ. Press.

Kringelbach, M. L. (2005). The human orbitofrontal cortex: Linking reward to hedonic experience. *Nature Reviews: Neuroscience, 6*, September 2005, P. 691.

Kringelbach, M. L., & Berridge, K. C.(2009). Towards a functional neuroanatomy of pleasure and happiness. *Trends in Cognitive Sciences, 13*(11), 479-487. doi:10.1016/j.tics.2009.08.006

Laundergan, J. C. (1982). *Easy does it*. Minneapolis, MN: Hazelden.

LeDoux, J. (1997).*The emotional brain: The mysterious underpinnings of emotional life*. NY: Simon & Schuster

LeDoux, J. (2002). *The synaptic self: How our brains become who we are*. NY: Viking.

Lewis, B. & Pucelik, F. (1990). *Magic of NLP demystified*. Portland, OR: Metamorphous Press.

Mann, K., Hermann, D., & Heinz, A. (2000). One hundred years of alcoholism: The twentieth century. *Alcohol and Alcoholism, 35*(1), 10-15.

Maslow, A. (1970). *Religions, values, and peak experiences*. NY: The Viking Press.

McClure, S. M., Daw, N. D., & Montague, P. R. (2003). A computational substrate for incentive salience. *Trends in Neuroscience. 26*(8), 423-8.

Miller, G. (1956). The magical number seven, plus or minus two. *The Psychological Review, 63*, 81-97.

NIDA. (2002). *Stress and substance abuse: A special report*. National Institute on Drug Abuse (NIDA). http://www.drugabuse.gov/stressanddrugabuse.html.

O'Brien, C. P., & Gardner, E. L. (2005). Critical assessment of how to study addiction and its treatment: Human and non-human animal models. *Pharmacology & Therapeutics, 108*, 18 – 58

O'Connor, J. & Seymour, J. (1990). *Introducing NLP*. London: Element.

Peele, S. (1989). *Diseasing of america: Addiction treatment out of control*. Lexington, MA: Lexington Books.

Peele, S. & Brodsky, A. (1991). The *truth about recovery and addiction*. NY: Simon and Schuster.

Prochaska, J. O., Norcross, J. C., & DiClemente, C. C. (1994). *Changing for good*. NY: William Morrow.

Progoff, I. (1959). *Depth psychology and modern man*. NY: The Julian Press.

Robins, L. N., Davis, D. H., & Nurco, D. N. (1974). How permanent was vietnam drug addiction? *American Journal of Public Health. Supplement, 64,* December, 1974.

Robinson, T. E., & Berridge, K. C. (2001). Incentive-sensitization and Addiction. *Addiction, 96*(1).

Robinson, T. E. (2004). Addicted rats. *Science, 305*(951).

Ruden R. (1997). *The craving brain*. NY: Harper Collins.

Schultz, W., Dayan, P., & Montague, P. R. (1997). A neural substrate of prediction and reward. *Science, 275,* 1593-1599.

Shattuck, D. K. (1994). Mindfulness and metaphor in relapse prevention: an interview with G Alan Marlatt. *Journal of the American Dietetic Association.* 94(8).

Skinner, B. F. (1957). *Science and human behavior.* NY:Free Press.

Sternman, C. (1990). *Neuro Linguistic Programming in alcoholism treatment*. NY: The Haworth Press.

Tobler, P. N., Fiorillo, C. D., & Schultz, W. (2005). Adaptive coding of reward value by dopamine neurons. *Science, 307,* 1642 -1645.

Waelti, P., Dickenson, A., & Schults, W. (2001). Dopamine responses comply with basic assumptions of formal learning theory. *Nature, 412.* July 5, 2001, p. 43.

Zoja, L. (1990). *Drugs, addiction & initiation: The modern search for ritual.* Gloucester, MA: Sigo.

Introduction to Exercises 1-7: Finding States, Anchoring and Future Pacing

A basic goal of these exercises is to provide a set of experiences so powerfully rewarding and intuitively valuable that it fosters the growth of conscious choice and remains a preferred set of life strategies for the individual participant. The skills and experiences created are designed to be intuitively real, accessible in every context and capable of enhancing practical efficacy in the real world. Used for self-development, they provide tools for personal exploration, enhanced creativity and personal growth. Applied in the context of substance use problems they additionally provide a strong experiential base for outframing addictive compulsion and awakening real choice. In the context of anger management they have repeatedly shown themselves as being extraordinarily powerful.

At the outset, our overall purpose is to create a set of conditions that awakens the pattern of being-in-the-world that Jung described as individuation and Maslow described as self-actualization. These are fundamentally spiritual directions. They also include powerful experiences of self-efficacy, ecstasy and personal worth. In the exercises that follow, we have operationalized them using behavioral and cognitive tools based on current research into the nature of memory and subjective experience. In so doing we replicate the physiology of spiritual experience and allow the participant to provide appropriate contexts whether broadly ecological (knowing one's place in the universe) or in terms of some more traditional notion of spirituality (Gray, 2001, 2002, 2003; Jung, 1965, 1976; Maslow, 1971; Newberg et al., 2001; Peck, 1968).

This first set of exercises is rooted in recent studies of the nature of memory, experience and basic Pavlovian conditioning. The first several exercises are aimed at using the brain's own mechanisms for creating present-time experiences of remembered events. These neurophysiological processes are being clarified daily and consist in the accumulation of layers of sensory data that recreate the original internal context and felt sense of the experience (Bechara et al., 2000; Damasio,1999; Freeman, 1998).

More specifically we begin with a set of techniques created by Richard Bandler to modulate the valence and quality of felt experience. These patterns, called submodalities, reflect the subjective experience of sense data. They were discovered using subjective techniques in the mid-1970s and are now being seen to accord with the emerging neuro-scientific evidence about the nature of memory and subjective experience (Bandler, 1993; Bandler & Macdonald, 1987; Bechara et al., 2000; Damasio, 1999; Freeman, 1998).

Much of what we experience emotionally is coded in terms of internal experiences of sensory data. Our emotions and states of mind change depending upon the sensory qualities applied to them in imagination. When things look bright we often feel better as opposed to taking a dim view of the situation. Likewise, the posturally keyed "feeling down" can be contrasted with holding one's head up. This translates into a subjective syntax of emotion and value that we will use to enhance the quality of the states chosen in the last exercise (Andreas, C., & Andreas, S., 1989; Andreas, S. & Andreas, C., 1987; Bandler, 1985, 1993; Bandler & MacDonald, 1987; Bodenhammer & Hall, 1998, Brookes, 1995; Dilts, 1993; Dilts, Delozier, Bandler & Grinder, 1980; Dilts, Delozier, Judith, A., & Delozier, Judith, 2000; Gray, 2001).

Research from standard psychological and neurophysiological sources has validated many of the visual submodalities. Fearful stimuli evoke different responses depending upon their distance or perceived distance. When the feared stimulus is far off, it evokes freezing responses that may allow us to avoid detection or assess the next action. Closer to the feared object, we run away and still closer we fight. Stimuli that seem to be moving toward us evoke more powerful responses, while those that are receding evoke lessened responses (Blanchard, Blanchard, Takahashi & Kelly, 1977; Muhlberger, Neumann, Wieser & Pauli, 2008). Moving stimuli, whether the movement is congruent with the expected movement of the object or not, the simple fact of movement awakens stronger emotions than do static stimuli (Simons, Detenber, Reiss & Shults, 2000; Simons, Detenber, Roedema & Reiss, 1999).

De Cesarei & Codispoti (De Cesarei & Codispoti, 2006; Codispoti & De Cesarei, 2007), showed that larger emotional stimuli evoked stronger responses than did smaller, independent of their valence. The same authors (De Cesarei & Codispoti, 2008) have also shown that focus, or the availability of fine-grained detail, affects emotional impact in the visual system. Pictures lacking fine-grained detail were perceived as less impactful than those containing high levels of detail. They also found that attention was based less on fine-grained detail than by whether there was enough detail to recognize the object.

Research into the functions of the orbito-frontal cortex, where the brain creates hierarchies of value, indicates that motivations reflected there are ordered preferentially in terms of the amount of detail that they provide and the richness of their representation across multiple sensory systems. Objects that are more fully represented across multiple sensory systems are perceived as more valuable or more threatening (Kringlebach, 2005). Further data emerging from studies of the superior colliculus--where spontaneous eye and head movements are controlled and where visual, somatosensory and auditory information is integrated--indicate that when auditory and visual impressions move together across the perceptual field, the neurons in that area fire more intensely. This has the effect of increasing the amount of attention paid to the object in question (Sparks, 1999).

There is a simple logic that suffuses the first three exercises: We use the brain's own multi-sensory strategy to enhance the felt sense of chosen memories. Having revivified a memory we use the same techniques to intensify the felt experience of the memory. Once the feeling has been intensified to ecstatic levels we use the limited capacity of short term memory (Miller, 1956) to turn the focus of attention to an ecstatic, floating experience that reflects some of the qualities reported by ecstatics (D'Aquili & Newberg, 1998, 2000; Laski, 1961; Maslow, 1971; Newberg, D'Aquili & Rause, 2000). Having awakened this ecstatic experience we then create a simple conditioned response that makes the resource available and malleable in any context. In the world of Neuro-Linguistic Programming (NLP) this is called "anchoring" (Andreas, C., & Andreas, S., 1989; Andreas, S. & Andreas, C., 1987; Bandler, 1985, 1993; Bandler & MacDonald, 1987; Bodenhammer & Hall, 1998; Brookes, 1995; Dilts, 1993; Dilts, Delozier, Bandler & Grinder, 1980; Dilts, Delozier & Delozier, 2000; Gray, 2001).

Beginning with the second exercise we identify past experiences of five resource states: Focused Attention, Confidence, Good Decision Making, Optimal Learning, and Having Fun. Each of the exercises in this section creates specific competencies including experiences of choice, mood control, self-efficacy and self-esteem. The specific states used here are adapted from the work of Carmine Baffa (1997).

Although, on the surface, we are teaching five simple response states, we are in fact accomplishing all of the following:

1) Simple Behavioral Effects. The anchoring exercises provide affective tools for counteracting negative states. They comprise a behavioral tool set that can be used as simple conditioned stimuli in counter-conditioning paradigms and in more extensive desensitization paradigms (Bandler & Grinder, 1975; Gray, 2001, 2002, 2008; Wolpe, 1982).

2) State-dependent Reframing. By orienting the participants towards positive states of mind, making them available in new ways, and enhancing those states, participants become more likely to experience positive aspects of their past lives through state dependent recall effects. As a result, their present experience is susceptible to more positive interpretation (Le Doux, 1995, 1998, 2002; Rossi, 1986; Tulving, 2002).

3) Response Generalization. Once positive responses are learned and appropriately framed, we use specific techniques to foster generalization of the responses to other contexts (Skinner, 1957; Bandura, 1997).

4) Body Awareness. An essential part of the Program is learning to pay attention to the kinds and sequence of sensory responses that signal emotional and physical states. As a result, participants become more associated to their own physical reality (Bandler, 1993).

5) Affective Choice Training. Participants who learn the anchoring skills attain significant training in the process of choice. As a result, reactive patterns begin to give way to the possibility of conscious choice as a matter of response generalization (Gray 2001, 2002; 2008).

6) Positive Self-Efficacy. As participants become more expert at defining their own affective state they gain a sense of their own capacity for choice and self-control. Self-efficacy is generated at a fundamental feeling level that is linked to a personal experience of making effective choices (Bandura, 1997; Gray, 2001, 2002).

7) State Orientation Shift. As they continue to practice the states and other exercises, the participants become more fully oriented towards their own positive potential. Past experience becomes a source of inspiration for positive change and choice (Damasio, 1999).

8) Resistance Destroyer. In the process of learning the basic states, each participant begins to discover good feelings within. In each session, a strong effort is made to have each participant experience intense positive feelings that s/he has personally generated. As a matter of simple conditioning, the basic patterns attach positive feelings to the facilitators and tend to make the sessions inherently rewarding (Gray, 2001, 2002).

9) Awakening the Choosing Self. As a result of the synergistic interplay of personal experiences in the Program, participants become aware of a transcendent whole, or Self, which represents them on a deeper level. This 'Choosing Self' becomes a center for positive future action (Gray 1997b, 2001, 2002).

10) On a neurological level, these exercises are believed to engage the frontal areas responsible for choice, evaluation and conscious attention and are significant to the program of reawakening frontal function in non-addictive patterns (Craig, 2001; Feil et al., 2010).

Exercise One spends time allowing the participants to taste the possibilities of subjective choice. We ask them to pick a state of mind which they enjoyed or found intensely pleasurable. Having identified a memory, time is spent enhancing it to ecstatic levels. For this exercise, we invite the participants to choose any state. We warn them that most of the states used in the other exercises must be clean, sober and legal. For this exercise, however, anything goes. We are not asking; however, we still prefer clean, sober and legal. Our intentions are to give participants a taste of endogenous altered states and to connect the Program with positive feelings of success and confidence. That the state might be immoral or otherwise forbidden increases the chances that they will respond well to the submodality techniques offered and connect those positive feelings to the Program more generally. That they are free to choose any state provides a positive link with the Program. In this exercise, the emotional power of the state is intensified in such a way that it transcends the original memory context and becomes an intensely powerful state beyond any specific memory connection. The abstraction of the state from the memory that was used to access it eliminates any risk rooted in the choice of the behavior.

In Exercise Two, Finding Resource States, we begin by identifying specific instances of each of the five root states. Several elements are crucial.

- Each example must represent a specific event in time.
- The events must be legal and sober.
- The memory should focus on an event that continues to be positively remembered into the present.
- The participant should focus on the very best part–the best ten seconds— of the memory.
- The memory should recall a situation in which the effects were the result of active choice or positive action on the part of the subject; they should not be passive states.

Having identified the memory, the participants are asked to write a short descriptive paragraph about the experience. This helps to stabilize the experience in memory. In a group setting each participant may be asked to share their memory. This also stabilizes the example, ensures that the example is appropriate and properly focused, and provides possible examples for the other participants.

In the second half of the exercise, the participants are asked to close their eyes and re-experience the resource state. They are asked to give it a name and to write the name down, further stabilizing the event. Once again the participants are asked to close their eyes and access the memory. This time, however, they are asked to notice how their senses are involved in the memory, how quickly they can access it and how much more detail is available. They are made aware of the memory enhancing effects of the exercise. Participants are to notice as much as possible the order of sensory information that they experience as they access the memory and its associated feelings. Facilitators can assist by demanding very specific sensory-based information. Is there a feeling? Where do you feel it? What are its qualities? Is there movement or temperature? Do they change? Using these kinds of questions, the participants begin to specify in real sensory terms their personal experience of the memory. Finally, they are asked to use the techniques from exercise one to enhance the felt experience of each memory to an ecstatic level.

Exercise Three, Anchoring Resource States, represents the actual anchoring process. This is a simple exercise in classical conditioning that provides each participant with an experience of their own ability to control their own feeling state. While NLP allows for a range of anchoring techniques, the Program relies on a well-established, classical conditioning paradigm. More specifically this is a modified delayed conditioning paradigm in which the onset of the unconditioned stimulus precedes the onset of the neutral stimulus (Bouton & Moody, 2004; Dilts & Delozier 2000; Grinder and Bandler, 1979; Linden & Perutz, 1998; O'Connor & Seymour, 1999; Rescorla 1988).

There are six things that we are seeking to accomplish:

- We are teaching the participants how to access and control positive emotional and physiological states.
- We are providing an explicit understanding of the nature and operation of triggers which places them expressly under the participants' control.
- We are enhancing their ability to control their own emotions in every context.
- We are building a sense of self-efficacy (most of us are taught to be controlled by our emotional states, not to run them for our own benefit).
- We are installing a powerful state of openness and teachability that can be used wherever the participants need it; and
- All of the states are connected to the presenter and virtually guarantee interest throughout the Program.

By this point, all of the participants have identified five resource states. They have spent some time analyzing them and noting the sequence of sensory information as they access them. They have also begun to notice that with each iteration, more and more of the content of the memory becomes available.

Using these same five states (Focus, Solid, Good, Fun, Yes) the participants are led through a simple conditioning paradigm that consists of accessing the memory, enhancing the experience of the resource state associated with the memory, attaching the resource state to a conditioned stimulus (a gesture) and enhancing the experience of the conditioned response state.

In the previous exercises we have specified and stabilized the memories upon which states are based. Now, we proceed to connect them to some meaningless gestures. We are careful to emphasize that these are meaningless gestures. For convenience sake we have chosen five simple gestures that run, in order, down the first three fingers of either hand: Tip of thumb to tip of index finger, tip of thumb to first joint of index finger, tip of thumb to tip of middle finger, tip of thumb to first joint of middle finger, tip of thumb to tip of ring finger. These gestures are illustrated several times, and each time the participants are reminded to use these gestures. Other gestures will work, but later, when we check for behavioral competency, we will look for these gestures.

The anchoring process is simple.

- Close your eyes and access the state. As you access it notice how it grows. As it continues to increase, and before it levels off, shake free of it and return to normal consciousness.
- Close your eyes and access the same state again. Notice the speed with which the state returns. Notice that it is stronger. As you notice this increase in intensity, make the approved gesture.

Continue to hold the gesture for about two seconds but only while the intensity continues to increase. End the gesture before the experience peaks. When the feeling has leveled off, shake free of it and return to normal consciousness.

- Continue repeating step two until you notice that the intensity of the experience increases when you make the gesture.
- When you notice that the gesture is enhancing your experience of the state, begin to pump or rub the contact point so that each pump of the gesture corresponds with the increase that the gesture is creating.
- Continue step four until the experience becomes very intense.

As simple as the conditioning process may be, it is important to realize that this is one of the places where the Brooklyn Program makes a distinctive contribution. There are many techniques for inducing altered states that have a certain utility in reducing cravings and changing mood. Yoga, acupuncture and Transcendental Meditation have often been used to good effect in the short term but, their utility often fades outside of the treatment context (Margolin et al., 2002; Morel, 1996). This may be because they are relatively context bound. Yoga (Hatha Yoga) depends upon positions and breathing patterns, Acupuncture depends upon the needles and a practitioner, while TM usually requires a focused state of internal directedness. All are linked to contextually bound conditioned stimuli.

The Brooklyn Program overcomes this problem by making the felt states available quickly, easily, unobtrusively and with an immediacy that can be surprising. By linking altered states to conditioned stimuli the behaviors become transportable. The conditioning emphasis and the later exercises that seed generalization into other contexts are intentionally designed to make the behaviors as available as a gesture. One participant dubbed the Program "Yoga to Go".

Exercise Four, The Keys to Enhancing Subjective Experience, reviews anchoring and provides multiple techniques for increasing the intensity of the anchored experience. In present iterations of the program it serves as a resource for the facilitator providing specific techniques for enhancing the states and the process.

Exercise Five, Getting to NOW, uses the technique of stacking anchors to assemble the initial five resource states into a sixth state that was not necessarily predictable from the others. NOW, the new state ,begins to explicitly constellate a positive state of Self as the emergent property of the interaction of the others. We frame the state as one that is useful for learning, choice and meditation.

The reframe itself is suggested by the classic experiment by Daniel Schacter and Jerome Singer (1962) in which participants' perceived responses to adrenaline were determined by information gleaned from context. Our understanding of the NOW state is that it represents a state of generalized autonomic activation. The state carries no intrinsic meaning except for that provided by the positive frame that has been previously established. Through all of the anchor-related exercises we continue to emphasize its flexibility general utility and deep, peaceful and ecstatic qualities.

Exercise Six, Pacing the Future, is an exercise designed to connect the experience of the positive resources to real life situations. It uses a technique of personal appointments through the day to remember to use and practice the states. The specific value of the exercise is that it seeds generalization of the new options learned in the Program to other contexts. One of the great faults of otherwise powerful therapeutic ideas is their failure to move change from the therapist's office to the real world. This exercise provides a crucial bridge between the Program sessions and the world of real life experience (Bandler & Grinder, 1976; Erickson & Rossi, 1980).

Exercise Seven, Taking Control, requires participants to create their own anchors. By now, each participant has had significant experience in creating and using the assigned anchors. Now we ask that they design and create a series of custom anchors. Just as the previous exercise bridges the session and the real world, this exercise removes control from the Facilitator and plants it firmly in the hands of the participants. This emphasizes the dual goals of Self-efficacy and Self-esteem.

The Exercises

Exercise 1
Feeling Good

Presuppositions Underlying the Exercise:

This exercise is intended to begin the process of orienting the participants towards positive resources. It sets up the basic premises of the Program in terms of present time experience of powerful endogenous states. It begins by challenging the participant to choose a pleasurable state without regard to its source or legality. By so doing, many participants will find the choice of a state easy as it allows them to indulge rebellious or negative impulses. Whatever the state chosen, the participant is led through a series of exercises illustrating that all emotions are subject to conscious manipulation. Because cravings and negative urges are emotions, the relevance of the exercise cannot be overestimated (Franken, 2003). Moreover, as the state is enhanced, the actual feeling is abstracted from the original memory context. As the feeling increases in intensity the memory fades away. The feeling tends to be transformed from something that *happens* to them to something their brain *can do*.

Repeated access to the memory provides practice effects for the positive feelings. There is usually a surprising experience of memory enhancement.

From the outset, it is the purpose of the exercises to enlist the participant in a series of pleasurable experiences which, superficially, have no relationship to drugs or treatment. The most important lesson here is that people can choose to feel better and there are simple techniques available to make that possible.

Certain of the root presuppositions of the wholeness perspective as it manifests in NLP are that people are not fundamentally broken and that every person has the resources necessary to accomplish their goals. This exercise begins to orient the individual towards that wholeness and to make them aware of their capacity to awaken resources that may be lying dormant within. It presupposes that people have access to memories that can be used as behavioral resources in the present (Cade & O'Hanlon, 1993, Erickson, 1954; Grinder & Bandler. 1979; Gray, 1997; 2001; Andreas & Andreas, 1987; Andreas & Andreas, 1989; Bodenhamer & Hall, 1998; Bandler & Grinder, 1975; Dilts, et al., 1980; Linden & Perutz, 1998; Robbins, 1986).

At this level, and continuing through the anchoring exercises, we assume Miller's (1956) discovery that the working memory store (short term memory) has a limited capacity. By emphasizing more and more features of the felt experience of the memory, we gradually abstract a feeling tone from the memory and allow the memory content to fade away. The "magic number" suggests that as more and more features of the feeling itself are emphasized, the content and context of the memory drop away. One of the indicia of success in this exercise and those following is access to a point where the state is no longer identified with a memory or memory context, but floats freely in a tranquil nether land associated only with the participant's awareness of the feeling.

The simple act of choosing a memory and manipulating the memory provides a powerful experience of self-efficacy. This serves to convince the participant that something interesting or valuable will follow as time goes on.

Expected Outcome

At the end of the exercise the participants will have identified a specific pleasurable memory. They will have accessed it several times and noted the difference between associated and disassociated states. They will have the experience of manipulating the sensory components of the memory and their own capacity to access a powerful, pleasurable resource state. In the process they will have discovered their own ability to enhance the memory.

Participants will begin to learn to describe their resource states and the associated affect in specific, sensory-based language that details an experienced sequence of visual, auditory, kinesthetic, olfactory and gustatory data.

Participants will gain an appreciation for the individuality of experience from person to person. They will have gained experience in moving into and out of positive altered states.

Instructional Notes

Throughout the exercise it is important to emphasize the foundational nature of these skills and that the participants are expected to follow the instructions provided. Remind them that even though, for now, they can choose any state they will, in the future, we will ask them to use very specific states of mind and that those states must be clean, legal and positive.

The aim of increasing self-efficacy is a crucial piece and needs to be emphasized throughout (Here is something that you probably didn't know that you could do. Try this, it will surprise you. Did you know you could do that?). It is also useful to emphasize to the participants that they are learning to use their brains in new kinds of ways.

There is a significant level to which this exercise provides an experience of hope. Participants who successfully complete the exercise have reason to believe that they know how to feel good or better and that they may be able to repeat it in the future.

There will be comments from time to time about how participants already use past memories to change their own mood. In the first part of the exercise, point out that daydreaming is a rather haphazard process. Here, using the submodality structure, we are learning how to use the brain's own processes to systematically and intentionally enhance memory. Indicate to the clients, that these exercises are designed to sharpen their memories by teaching them to return to the same memory at will and to be able to focus on any part of it. Remind them that we will be using the present skills to create buttons for automatic access to altered states. Recall how, from time to time, it can be difficult to enter an appropriate state of mind and point out that we are going to make state change as simple as making a gesture.

Invite the participants to choose an experience that made them feel wonderful. It may have been empowering, fulfilling, fun or ecstatic. Do not ask them to tell you what the experience is. Let them keep it private. The important issue in the first exercise is that they gain some experience with the techniques. However, in the first exercise we emphasize the following criteria for choosing remembered states:

1) Choose one specific moment in time (not a series of times). This might be experienced as a short movie or still picture, ending at the most intense part of the experience.

2) The memory should be emotionally clean. It should not (intrinsically) carry the emotional baggage of regret or bad circumstances.

3) The memory or circumstance should be stable over time and not subject to transformation (such as focusing on a present job or relationship that could be lost or destroyed).

4) All examples should be experienced for themselves, without regrets or negative baggage. If a state cannot be used without self-pity or remorse, another state must be used. If a participant refuses to access states without making such attachments, they should be invited to work one-on-one at a later time.

Suggest that childhood memories of innocent experiences are just fine. Remind participants just to go for the memory in isolation. For all examples, just get into the memory for the sake of how it felt then.

It is important to suggest that the example need not be the best thing that they ever experienced; a good memory is fine.

In the first exercise we do not specify the state and ask the participants to not describe the state, unless they have troubles later in the exercise. This means that some of our participants will use illegal, immoral and otherwise objectionable states. This is not a problem. The point of the first exercise is to engage participation and to ensure that each participant has an experience of the positive impact of the technique. If the exercise is being done properly, the memory with which they began will fade as they become absorbed in their own experience of control and ecstasy. In effect, the techniques transform the experience from a rebellious response into one that serves the purposes of the Program. From the perspective of classical learning theory, the last experience in the sequence, an experience of efficacy, sober ecstasy and personal empowerment is the response that will be reinforced (Thorndike, 1911).

An important part of Exercise One is an effort to familiarize the participant with the elements of emotional experience. These elements are referred to in the literature of NLP as submodalities.

As they access the states, they are asked to notice the difference between associated and dissociated experience—in the picture or out of the picture. They vary the intensity, — bring it closer, make it brighter, make it louder. After each change they are asked to note the change in felt experience. Each instruction is designed to provide a felt change in the experience and to provide practice in the manipulation of feeling by changing the submodality qualities of the experience (Andreas & Andreas, 1987; Andreas & Andreas, 1989; Bandler & Grinder, 1975; Bandler & MacDonald, 1987; Bodenhamer & Hall, 1998; Dilts, et al., 1980; Gray, 2001; Grinder & Bandler, 1979; Linden & Perutz, 1998).

It is important to emphasize that not every remembered experience will have the impact of a photographic memory. Initial experiences are often weak and must be enhanced. This is the specific value of the submodality manipulations; a systematic means for controlling the valence and intensity of subjective experience. At the outset, whatever sense of the memory is available will work well.

Have the participants close their eyes and experience the memory. Let them note just how they get to the memory and what they notice first: a picture, a smell, a sound or a feeling. What comes next and next and next? One client described his access to a time of focused attention as first hearing confused sounds, then having a feeling of butterflies in his stomach. This was followed by a sharp smell of specific odors associated with the incident and another increase in feeling. He next found himself focusing on the face of someone and a further intensification of the feeling.

Read through the Script in a normal, instructional tone and give the participants an opportunity to adjust their perceptions accordingly. Remind them that the process is easy and that imagination makes it work best.

Advise the participants that your suggestions are just that— suggestions that they can try. If there is no picture at first, turn to the sound. If there is only feeling, stay with it and don't worry about the other parts. Reassure the participants that whatever sensory manipulation that they can use to enhance the feeling is just right.

In one of our groups, a color-blind participant asked what he should do with the instruction to turn up the vividness of the colors. He was advised to notice what would happen if he could. He reported an immediate increase in the intensity of the experience.

One of the important techniques used here is the observation of the way an emotion spreads through the body. Richard Bandler, the originator of most of these techniques, points out the value of paying attention to these patterns and learning to speed them up to enhance the feeling (Bandler & MacDonald, 1987; Bandler, 2000).

As each person goes through the exercise, have them carefully note which sensory system impacts their experience the most.

After accessing the resource state several times, begin to suggest that they can now "zoom" into the very best part; the place they left off at the last access. After a quick visit with the memory, have them come fully into the present and shake out the state. Have them literally shake their bodies to reorient them to the present. Now suggest that they return to the same state, noting

    A)   Where the feeling starts.

B) How and where it spreads in their bodies.

C) How it reaches peak; and

D) How it leaves their body.

Again, have them come fully into the present and shake out the state. Now have them return to the state and quickly enter into the feeling state. As they note the rush of onset (call it a "rush") as the memory reasserts itself, have them draw the energy back to the starting point so that the experience feeds forward through the cycle, increasing in intensity. In the exercise we use the phrase "using imaginary hands." This reflects recent neurophysiological research that shows imagined physical manipulations enhance the capacity to manipulate imagined objects (Downs et al., 2002).

The following language may be useful: "Imagine that you can reach out with imaginary hands and take hold of the best part of the feeling as it spreads through your body. Take hold of it and bring it back to the place where it started. Push it back through the center so that it doubles. Continue to push it out through your body and notice how it grows stronger. Grab the best part again and push it back through the center." Repeat this cycling, faster and faster until the state becomes surprisingly powerful. Remind participants to attend to the cycling of the feelings not the picture. It can also be useful to think of stirring or turning the felt sense (Bandler, 2000).

As noted in the earlier introduction, remind the participants to focus more and more on the qualities of the felt state. Overload short term memory with impossible dimensions of feeling: location, texture, how it spreads, depth, breadth, height, temperature, imagined color and imagined sound. As the participants focus on more and more of these, the context and content will be crowded out of working memory and they will be left in a powerful, peaceful ecstasy. It is a generalized state of autonomic arousal that is framed by the original state.

An important part of the exercise is the abstraction of the feeling from the memory. We begin with a remembered experience to gain access to a feeling state. We enhance the memory to increase the felt sense of the experience. We then focus more and more on the feeling in order to lose the connection with the memory and discover the feeling as something associated with the participant's own capacity to feel; independent of external influences.

In the end, all of the states should result in a tranquil, ecstatic or peaceful experience that may or may not carry the flavor of the original emotion but is completely disconnected from the memory itself. By abstracting it, we gain a completely transferable resource. By making it strongly pleasurable, we gain a motivation for practice, increased probability of use and a set of positive experiences that can compete with cravings.

Process Summary and Script

Use this script for the first several passes through exercise one. It may also be used anytime that clients are unsure of the process or unable to create the altered states.

1. Invite the participants to choose an experience that made them feel wonderful. It may have been empowering, fulfilling, fun or ecstatic. Let them keep it private. The aim of the first exercise is to gain experience with the techniques.

2. In the first exercise, *we do not specify the state* and do not ask the participants to describe the state, unless they have troubles later in the exercise. *This means that some of our participants will use illegal, immoral and otherwise objectionable states. This is not a problem.*

3. Have the participants close their eyes and experience the memory. Let them note just how they get to the memory: what they notice first, a picture, a smell, a feeling? What comes next and next and next?

4. As they access the states, they are asked to notice the difference between associated and dissociated experience—in the picture or out of the picture. They vary the intensity, — bring it closer, make it brighter, make it louder. After each change or set of changes (see below) they are asked to note the change in their felt experience. Each instruction is designed to provide a felt change in the experience and to provide practice in the manipulation of feeling by changing the submodality qualities of the experience.

5. During the initial walk-throughs, as you are reading the submodality lists, make an effort to use normal intonation and volume, with no effort at trance language or special emphases. Persons in substance use treatment are notoriously paranoid and can respond badly to unconscious communication styles if used prematurely. Use trance patterns only after they have created altered states for themselves. These will be treated later in the training.

   - **People differ as to which sense arises first when they access a memory. Some people remember pictures; some sounds; some begin with feelings. Most people in the West prefer vision. For this reason, we are starting with the visual part of the memory. If you find that sound or feeling comes up first for you, feel free to start there and return to the other senses in the way that works best for you. But please read through the rest of the exercise before starting.**

   - **One more thing; All of this is easy. Most of it consists in just noticing how things are. The simple act of turning your attention to the sensory distinction is often enough to change it. In every other case, gentle imagining works fine.**

Script

For Exercise One have the participants choose a state using the criteria noted above.

Have the participants close their eyes and experience the memory. Let them note just how they get to the memory: what they notice first, a picture, a smell, a feeling? What comes next and next and next?

Use the bold language that follows as a script.

Notice whether, in your imagination, you are experiencing the memory from within, or watching it from outside like in a movie.

If your memory seems to be just in your head, imagine that you can *step all of the way into it*. As you experience the memory, you may even notice flashes that feel like really being there, focus on these. Take a few minutes to make sure that you are actually in the experience. Once you have the sense of really being there, even if it was only for flashes, come fully back into the present context.

Once you have a sense of what it's like to relive the memory from within, step all the way back into it and get a feel for it. Notice that you can step right into one of those parts where it all came alive. Step right into it. Notice what you are seeing and feeling and hearing. Notice the patterns of tension in your muscles. Notice who is there and how you feel emotionally. Take a few minutes to get really familiar with the feel of being there. Enjoy it. Come fully back into the present.

<u>From this point forward drop the use of the word memory and begin to refer to</u> *<u>the feeling, the experience or the resource.</u>*

Once the client has stepped into the experience, they can then begin to vary the submodality structure of the memory. Instruct them to make the changes in a way that makes the experience work best for them. Let them experiment with each dimension to find a level that feels best.

Go through these one at a time, pause after each to allow for their processing time and, in the early exercises, ask them to describe how each change affects the experience. Tell them to remember the ones that work best.

After each change, ask them to note the change in their felt experience. Each instruction is designed to provide a felt change in the experience and to provide practice in the manipulation of feeling by changing the submodality qualities of the experience. Remind them to take note of the kinds of perceptual changes that make the most positive difference in the experience.

Once more, step all the way back into the experience and get a feel for it. Notice that you can step right into one of those parts where it all came alive. Step right into it. Imagine that you can bring the picture closer and make it bigger and brighter.

Each of you has had the experience of changing the size of a picture on a computer screen and changing the brightness on a television set. Use your imagination, now to make the image closer, bigger and brighter. Notice how that changes your experience. Keep the change that produces the best experience. If the experience lessens put the experience back the way it was.

Notice whether the image is in color or black and white—if it is black and white, turn on the color. If it is already in color, turn up the intensity. Notice how that changes the experience. Find out how much color feels best.

Notice whether the experience is moving or still. If it is still, turn on the motion. If it is moving, turn it into a still image. Notice which one feels best. Keep the change that makes the experience feel best.

Notice whether the sound is on or off. If the sound is off, turn it on. Adjust the volume of the sound so that it enhances the experience.

Notice how the sound moves. Notice how the sound moves with the experience. Notice whether the sounds are noises or music or voices. Pay attention to where they come from.

Breathe in through your nose and smell the smells that were present there.

Come back for a moment, shake out the experience and talk to me about what happened. Did that feel good? Did you know that you could do that?

What worked best for you?

> After a few minutes of discussion invite them to just close their eyes and return to the place where they left off and continue as follows.

Now, step all the way back to the point where you just left off. For some of you the memory has gone away and you were just out there floating, that's good, go back there. Go back to the state where you left off and notice how easy this is.

Notice how you breathe in this experience. Notice how you hold your body --the patterns of tension and relaxation that enhance your experience. Adjust your posture, so that it enhances the experience.

Notice how you breathe in this experience and the expression on your face. Adjust your expression so that it enhances your experience.

Return to the experience and zoom right back to the very best part. Turn up the brightness, bring it closer and turn up the volume on the sound. While you do these things, note the path of the energy through your body. As you notice the feeling getting stronger, loop the feeling back through the starting point so that it doubles up as it moves through you. Notice that it moves further, faster and more powerfully.

Pay attention to how that works. Find out how many ways you can enjoy it.

> (Provide a few minutes for self-exploration)

And now come all of the way back.

> Now have them return to the state and quickly enter into the feeling state. As they note the rush of onset (call it a "rush") as the experience reasserts itself, have them draw the energy back to the starting point so that the experience feeds forward through the cycle, increasing in intensity. In the exercise we use the phrase "using imaginary hands."

The following language may be useful:

[LOOP}

Once again, step right back into the place where you left off and feel the rush of feeling as you step back in.

Imagine that you can reach out with imaginary hands and take hold of the best part of the feeling as it spreads through your body.

Take hold of it and bring it back to the place where it started. Push it back through the center so that it doubles.

Continue to push it out through your body and notice how it grows stronger. Grab the best part again and push it back through the center.

Repeat this cycling, faster and faster until the state becomes surprisingly powerful.

- Remind participants to attend to the cycling of the feelings not the picture. It can also be useful to think of stirring or turning the felt sense.
- Remind the participants to focus more and more on the qualities of the felt state.

Overload short term memory with impossible dimensions of feeling: location, texture, spread, depth, breadth, height, temperature, imagined color and imagined sound. As the participants focus on more and more of these, the context and content will be crowded out of working memory and they will be left in a powerful, peaceful ecstasy that carries the flavor and physical tone of the original state. It is a generalized state of autonomic arousal that is framed by the original state.

…And as you turn your attention, …just gently turn your attention, … to the center of the feeling, you can begin to notice, … really notice… its temperature, … its color…. Notice whether it makes a sound , … or a hum. And you can notice, really notice,… how the feeling moves…. Whether it is centered in your body, or beyond your body…. Whether it moves in a circle … or a loop, … or a spiral…..whether it turns clockwise or counterclockwise … and whether it turns like a wheel …or like a turntable…. And as you notice the pattern of this movement, … you can reach out with imaginary hands … and begin to trace this movement… with those imaginary hands, … and if the movement of the feeling …is not a complete movement, … you can take those imaginary hands … and guide that feeling … through its own pattern, … back into its own center, … so that it grows …. and increases … and flows and multiplies. … And you can use those imaginary hands … to take hold of the feeling …. and move it faster … and faster … through its own center so that it doubles … and doubles again, … and grows stronger … and stronger, … and the pictures fade, … and the memories fade … and you find yourself floating … and resting, … down, … all … the … way … down, …into pleasant, …. safe and ….warm. …Resting ….into your own ability … to feel …. good … now….

Allow participants to remain in state for a while. They may safely be allowed to remain in this state for extended periods.
Gently call the participants back to the present time and place.

Come on back. Reorient to the room and the present context in a way that is comfortable and that allows you to retain the lessons of this exercise in the present context. Come on back. NOW.

An important part of the exercise is the abstraction of the feeling from the memory.

- We begin with a remembered experience to gain access to a feeling state.
- We enhance the memory to increase the felt sense of the experience.
- We then focus more and more on the feeling in order to lose it from the memory and discover the feeling as something associated with the participant's own capacity to feel; independent of external influences.

After a brief discussion, repeat the sequence Beginning at the section labeled [LOOP], adding the words

Step back into the state, zoom right back into the very best part and discover how much you can enjoy that state and how many dimensions of wonder you can find inside."

At the end of the session, participants are sent home to practice the enhancement techniques with several more experiences of their own choosing. In practice, these will be reviewed and re-accessed at the beginning of the next meeting. This will begin the session with a positive bias towards the facilitator and more generally towards the techniques.

Behavioral Standards

For this exercise, each participant will be able to name and access a series of memories which they have manipulated to ecstatic levels. They will form a basic understanding of the submodality distinctions and a budding appreciation of their own capacity to change their own mood.

Sequence Note:

You may begin to use the techniques from exercise 4: Keys to Enhancing Subjective experience at almost any time.

Exercise 1
Feeling Good

Recent research (Damasio, 1999; Freeman, 1998) has begun to show that memories are not so much stored in our heads as they are reconstructed. Antonio Damasio (1999) points out that the brain has no direct experience of the world. What we experience is our bodies' response to sensory stimulation. So, each experience that enters awareness is a sequence of physical, neuronal and hormonal adjustments and includes things like changes in posture, changes in eye focus, dilation or contraction of the pupils, variations in the tensions of the muscles in the inner ear, flaring of the nostrils, adjustment of facial muscles and remapping of remembered experiences in the original sensory areas (Bechara, et al., 2000). Each memory experience is created new from this complex interaction of nerves, hormones, muscles and physical movements as they replicate the physiology of the original perception.

The brain uses multiple sensory systems to build up memories and present-time experiences, layer-by-layer. As the details from multiple sensory systems come together, a unifed, present-time experience of the object of attention arises into consciousness. Francisco Varela (et al., 1991) has estimated that only 10% of the information processed by the brain represents direct sensory input. The remainder is interpretative feedback from the rest of the brain.

If we start with a remembered image, a map of the retinal image (at the back of the eye) is transmitted to the primary visual cortex (in the back of the head). After the image registers, feature detectors combine with inputs from short term memory (where we experience attention) to recognize the basic form of the pattern or thing seen. The brain now starts to amplify the pattern while the rest falls into the background. As the pattern gains clarity, it begins to awaken connections to other sensory systems. These feed back into the original perception, strengthen it, clarify it, and bring in connections to information from other sensory systems. This pattern of activity continues until enough information is added so that context (the meaning provided by our surroundings), is added from the hippocampus. This contextual information feeds back through the entire loop fostering the addition of still more data until the whole is clear enough to awaken the emotions and feelings that originally accompanied the memory. The memory is reconstructed in present time as the brain weaves together multiple layers of sensory information into a recognizable experience.

Neuro-Linguistic Programming (NLP) has taken this process and developed a way to systematically use sensory information to create present time experiences of remembered events. By carefully mapping, organizing and adjusting the sensory information from a memory and systematically going through each sense with its submodalities, NLP uses the brain's own rules to enhance and recreate remembered experience. In NLP, each sense is described as possessing several submodality features. These include things like location, intensity, distance, and dimension. We will discuss the individual submodalities in depth below. The roots of this exercise are to be found in the works of Richard Bandler (Bandler, 1993, Bandler & Macdonald, 1987)

What makes submodality distinctions important is their ability to change present time experience. Submodalities represent the brain's control system for the way we feel. Just as the zoom button on a camera changes the size of the picture, and the volume knob controls the loudness of a sound recording, changes in the submodality structure of an experience change the meaning and intensity of an experience. By changing submodalities, you can start with the shadow of a memory – the memory that something happened – and end with a real-time felt experience of the memory. Further, submodality manipulations will allow you to find the emotion in the memory and enhance it separately.

This means that, for the following exercises, it does not matter how well or how poorly you remember an event. We will be teaching you how to use the brain's own control system to create a full experience of several memories.

People differ as to which sense arises first when they remember something. Some people remember pictures; some sounds; some begin with feelings. Most people in the West prefer vision. For this reason, we are starting with the visual part of the memory. If you find that sound or feeling comes up first for you, feel free to start there and return to the other senses in the way that works best for you. But please read through the rest of the exercise before starting.

One more thing; all of this is easy. Most of it consists of just noticing how things are. The simple act of turning your attention to the sensory distinction is often enough to change it. In every other case, gentle imagining works fine (how would this look or feel or sound if I made this change?).

Start off with a pleasant memory; a memory that was fun or interesting or positively moving. Go for a memory that stands on its own as something pleasant. It doesn't have to be the best thing in the world, just something you enjoyed. Go through the list. Note anything that intensifies your experience of the memory. Fiddle with some of the possibilities and see what feels best. When you find something that dramatically enhances the experience, keep the change. In the lists that follow we will start by listing the possibilities and continue by inviting you to play with them.

Vision has the following features or submodalities:

Association: Are you experiencing the memory from the perspective of an actor in the scene or as an external observer? Can you see it through your own eyes? Does it seem like you are watching a movie, or are you in the action?

Color: Is the experience in color or black and white?

Brightness: Is the experience brightly or dimly lit?

Focus: Is the image focused or unfocused?

Frame: Is the experience framed or unframed?

Aspect ratio: Is the experience tall or short: wide or narrow?

Dimensions: Does it have one dimension or two or more dimensions?

Movement: Is it a movie or a still picture?

Distance: Is it near or far?

Size: How much of the visual field is filled by the image?

Now, go back through the list and make the suggested changes. If the change doesn't have an impact on your feelings, or it weakens the experience, put it back the way it was.

Association: Make sure that you're experiencing the memory from within. If you seem to be watching from outside, use your imagination and step all of the way into the experience. Notice what changes in your experience.

Color: Notice whether the experience is in color or black and white. If the image is black and white, use your imagination to turn on the color. If it is already in color, or if you've just turned the color on, turn up the intensity. Notice the difference.

Brightness: If the image is dim, turn up the brightness, — just enough to reveal more detail. Notice the difference in your experience.

Focus: Where do you focus your attention in the image? Is everything in focus? Can you change the focus? What focus gives the most impact?

Frame: If it is framed, remove the frame and make it panoramic. What changes?

Dimensions: If the experience has two dimensions, imagine three dimensions. Extend them into the plane. Add a sense of time or eternity. What happens to the experience?

Movement: If the representation is a still picture, make it a movie. Note the change.

Distance: Bring the picture much closer. How does the impact change?

Size: Make the picture much larger. Double it and double it again. What happens?

Auditory submodalities can add significant depth to an experience. Some of the more significant auditory submodality distinctions are as follows:

Volume: How loud is the sound?

Sources: Are there one or more sources of the sound?

Dimension: Are the sounds mono, stereo or holophonic?

Direction: From what directions do the sounds come from?

Type: What kinds of sounds are there? Voices, music, just sound?

Timbre: How rich or complex is the sound?

Rhythm: What rhythms are there in the sounds?

Return to the same memory and make the following adjustments and observations. Again, notice which changes carry the most impact.

Volume: Make the sound louder. Turn up the volume. Adjust the volume for the maximum positive impact.

Sources: Notice where the sounds come from.

Dimension: If the sounds are monophonic add stereo or holophonic sound.

Direction: Notice the directions of the sound sources. Pay special attention to the ones that move with objects that you see.

Type: Notice whether the sounds are voices, music or just sound.

Timbre: Note the richness and complexity of the sound. Does it resonate in your body?

Rhythm: Notice any rhythms in the sounds. Do they resonate in your body or move with any seen objects?

By now, you should have noticed that the experience that you began with has grown much stronger, much more vivid and more real. Any emotion associated with it should already be growing strong. You may have also realized that when you stepped back into it to manipulate the sound dimension, it was already easier to get into. The more sensory data that is added to the original experience, the stronger and the more detailed it becomes in consciousness. By now, you have already noticed something of the feel of the experience. Feeling — kinesthesia, —has the following dimensions:

Depth: Do you experience emotion, physical sensation or both?

Location: Where do you feel it? In one place or several?

Movement: Is it moving or still?

Dimension: How does it spread, one dimension, two dimensions, three or more?

Intensity: How strong is the feeling?

Texture: Is it smooth or rough, ragged or even?

Temperature: Cold, warm, hot, changing?

Moisture: Moist or dry?

Now, step back into the same experience. Enter quickly and enjoy the speed with which the it arrives. Notice the rush of sensory information. As you step into it, make the picture much larger, turn up the volume and pay close attention to how the experience arises in your body. As you enjoy these sensations you may even notice that you remember more detail from within the experience. Play with the following submodality distinctions. Notice how they change your experience and note which has the most impact. Stay with it for a while and enjoy it.

Depth: Do you experience emotion, physical sensation or both? If you are only feeling one, add in the other.

Location: Notice where in your body you feel the sensations. Where do the emotions start?

Movement: Notice if the feelings move. If they do, note where they start and how they spread. Notice where they are strongest and how they leave the body.

Dimension: When you experience a feeling or emotion and you notice that it spreads, notice how it spreads. Does it spread in one dimension as a line, two dimensions like a plane or disk, three dimensions like a ball, or more dimensions than you can express?

Intensity: Notice the intensity of the feeling. Double it and double it again. Adjust the level of intensity so that it becomes most pleasurable.

Texture: Just notice if the texture is smooth, wavy, rough, ragged or even. If one feels better, try it on.

Temperature: Is the feeling cold, warm or hot? Is it changing?

Moisture: Is the feeling moist or dry?

At this point, the intensity of the experience should be surprising. You have been working with the brain's own intensity controls and you may have noticed that you can do some amazing things with your own feelings.

Now, quickly, step back into the experience again. Notice the rush. Do it fast to maximize its intensity. As you enjoy the rush, turn up the sounds, enhance the colors, and make the picture bigger and closer. Notice where in your body the emotions begin and how they spread. As you become aware that the feelings are intensifying, notice where and how far they have spread. Imagine that you can grab all of that intensity and recycle it though the place where it started. Reach out with imaginary hands and draw that intensity back through its own center. Notice that as it flows out again, it is stronger, deeper, thicker, and it spreads further through your body. Catch it again and bring it back through the center so that it starts to spin through its own circuit and doubles with each loop. Keep it spinning until the image disappears and the room disappears and you find yourself floating in bliss.

When you come back – you might drift off or just return spontaneously, — come completely out of it and try it again. This time, though, you'll notice that all you have to do is turn your attention to the experience and the feeling should begin to arise. Turn your attention to the feeling. Let it start to spin and notice how quickly you can get someplace very interesting.

At this point you may notice that the memory itself just fades away. Let it happen. As you spend more and more attention on the feeling and your favorite parts of the feeling; as you spend more and more attention on how many levels of depth and peace, enjoyment and glory are wrapped up inside of you; the will gently fade from consideration.

When this happens, when the memory itself fades, but the feeling remains, you have crossed a subtle threshold. Emotion has begun to be something that you can *do*, not just something that happens to you. You have chosen to feel something and you now have subjective tools for doing it again. You can do it with any feeling that you have ever had.

Unfortunately, for humans, most of our practice with these tools has led us to enhance the wrong feelings. We have, in the past, used them to create anger, depression and shame. It is now time to use these tools to grow feelings of love, joy, tranquility, peace, hope and strength.

The following steps summarize the process we have just completed.

Find several exquisite experiences and magnify them one at a time using the following steps.

Please read each instruction completely before beginning

1. Think of a time when you felt wonderful.

2. Notice whether, in your imagination, you are experiencing the memory from within, or watching it from outside like in a movie.

3. If your memory seems to be just in your head, imagine that you can *step all of the way into it*. As you experience the memory, you may even notice flashes that feel like really being there, focus on these. Take a few minutes to make sure that you are actually in the experience. Once you have the sense of really being there, even if it was only for flashes, come fully back into the present context.

4. Once you have a sense of what it's like to relive the memory from within, step all the way into it and get a feel for it. Notice that you can step right into one of those parts where it all came alive. Step right into it. Notice what you are seeing and feeling and hearing. Notice the patterns of tension in your muscles. Notice who is there and how you feel emotionally. Take a few minutes to get really familiar with the feel of being there. Enjoy it. Shake it out of your body and come fully back into the present.

5. Step back into the experience. Again, notice how you can zoom right into the best part. As you do so, imagine that the image is 40 feet tall and 40 feet wide. Become aware of the sound and the directions from which the sounds come. Notice how these enhance the experience. Come fully back into the present.

6. Now, return to the experience once more. As you do, notice that you can zoom right to point where you left off the last time; right to the very most intense part. Make it bigger and brighter and closer. Turn up the volume of the sound. Notice the rush of feelings and sensations. Pay attention to the feelings and notice

   A. Where in your body does the feeling start?

   B. How does it spread through your body to peak intensity?

   C. How does it dissipate?

   Shake out the feeling and return to the present.

7. Return to the experience and zoom right back to the very best part. Turn up the brightness, bring it closer and turn up the volume on the sound. While you do these things, note the path of the energy through your body. As you notice the feeling getting stronger, loop the feeling back through the starting point so that it doubles up as it moves through you. Notice that it moves further, faster and more powerfully.

8. Continue to recycle the energy in this manner. Do it faster and faster, until you lose any sense of the image and find yourself floating, immersed in the feeling alone.

Further Applications

Beyond laying the groundwork for the other exercises, the submodality resources and techniques found in this exercise can be used for other purposes.

Memory Enhancement. Pick something that you would like to recall more clearly. Use the steps from this exercise to enhance it. When the feelings comes on, spin it just enough to enhance the experience, so it feels as if you are really there. Explore the memory and discover something that will surprise you or meet your needs.

Sometimes, just imagining that the visual part of the memory is very big–use imaginary hands to stretch it–can provide extraordinary levels of detail.

Sensory Enhancement. Imagine that you have huge ears. Imagine having ears three feet tall that you can turn in any direction. As you hold that image in your mind, notice what happens to your hearing. Do the same thing with your eyes or nose–any sensory organ —or body part.

Transforming Emotions. Use the submodalities in reverse to reduce the impact of an upsetting or fearful memory. Do the same to present time events. Make the picture small and distant, flatten it out and drain out the color. Turn down or distort the sound. Step out of the picture and watch it from a distance.

If you are upset by the memory of a critical voice, try speeding the voice up so that it sounds like Mickey Mouse. Put circus music behind the voice and discover how it changes. Perhaps you could make it very loud or very soft. Play with it (Bandler & Grinder, 1976).

References for the exercise

Bandler, R. (2001). *Design human engineering* (Audio).

Bandler, R. (1993). *Time for a change.* Capitola, CA: Meta Publications.

Bandler, R. & MacDonald, W. (1987). *An insider's guide to submodalities.* Moab, UT: Real People Press.

Bechara, A., Damasio, H., & Damasio, A. R. (2000). Emotion, decision making and the orbitofrontal cortex. *Cerebral Cortex, 10(3),* 295-307

Damasio, A. R. (1999). *The feeling of what happens: Body and emotion in the making of consciousness.* NY: Harcourt.

Freeman, W. J. (1998). The Neurobiology of multimodal sensory integration. *Integrative Physiological & Behavioral Science,* 33(2).

LeDoux, J. (2002). *The synaptic brain.* NY: Viking Penguin.

LeDoux, J. (1998). *The emotional brain.* NY:Touchstone.

Varela, F., Thompson, E., & Rosch, E. (1991). *The embodied mind: Cognitive science and human experience.* Cambridge: MIT Press.

Exercise 2
Finding Resource States

Presuppositions Underlying the Exercise:

This exercise is intended to continue the process of orienting the participants towards positive resources. It sets up selection criteria for acceptable and non-acceptable resource states and continues the analysis of the states into sequences of sensory data.

Repeated access to each memory will provide practice effects for the positive feelings. Participants often experience memory enhancement.

Many of the presuppositions upon which this exercise is based were reviewed in the previous chapter.

A major presupposition understands that positive affect is evoked as an idiosyncratic sequence of sensory images. Each memory is coded differently for each person. Further, each person will unconsciously select a series of images that will combine in later exercises to awaken a deeper sense of Self. The outworking of these choices in a group context illuminates the process for all the participants and helps to break down resistance.

The act of choosing a memory and manipulating the memory provides a rudimentary experience of Self-Efficacy.

Expected Outcome

At the end of the exercise the participants will have identified five individual resource states following the choice criteria in the lesson. In general, these will represent experiences of focused attention, the moment of decision, personal discovery, fun and confident competence.

Participants will differentiate between appropriate and inappropriate exemplars. Appropriate exemplars are positive in nature; they do not include intoxicants, criminal behavior or injury to another or another's property. The exemplars chosen should not be linked to substance abuse/dependence, incarceration, criminality or other negative experiences even if they represented significant high points of choice ("I decided not to use again."), emotional release ("I felt good when I got out of jail."), etc.

Participants will learn to describe their resource states in specific, sensory-based language that details an experienced sequence of visual, auditory, kinesthetic, olfactory and gustatory data.

Participants will gain an appreciation for the individuality of experience from person to person. They will have gained experience in moving into and out of positive altered states.

Instructional Notes

Throughout the exercise it is important to emphasize the foundational nature of these states and that the participants are expected to follow the instructions provided. Remind them that even though, for now, they are expected to follow our instructions; this is only to learn the basic skills. When they have learned these, they will have a skill that can be employed as they please.

The aim of increasing self-efficacy is a crucial piece and needs to be emphasized throughout. It is also useful to emphasize that the participants are learning to use their brains in new kinds of ways.

When used in the context of a group Program, It is our practice to complete about half of part one and half of part two in the first session and to leave the remainder to be done as homework. This ensures that the entire exercise is understood in session.

At the outset, the target states need to be explained.

FOCUS: This should be an example of focused attention. It might have been a time when you were watching an exciting movie or reading an interesting book. It might be a time when you were playing a game or doing something exciting. Whatever it was, it was a time when you were really THERE, time disappeared and you enjoyed it totally. It might be a time when you met someone new that you were very interested in getting to know. A time when you were able to spend hours with that person, but it only seemed like minutes. Make sure that it was something that you enjoyed. Choose a memory with no regrets or mixed feelings.

SOLID: A time when you made a good decision, one that continues to be satisfying even today. Maybe you bought a car or a house and it proved to be a good investment. For our purposes, find an example that involved a real choosing process. Find one that began with many possibilities but ended with a single choice. For example, think of going to buy clothes. Think of the initial choice of a store; then, of a brand or price range; then, of a style. At some point the choice narrows to just a few possibilities. As you make the final decision there is a feeling that tells you, "This is it." This should be an experience of *choice*, not of *settling*.

GOOD: A time when you totally surprised yourself by being able to learn a complex task and do it well, despite the fact that you didn't think that you would be able to do it. What might make this experience special is that it wasn't until you had already learned and were already doing the new thing that you finally realized you were doing it! I often think of trying to learn to ride a bicycle or drive a stick shift. There is that one minute when it all comes together. Find a moment like that, when something difficult suddenly comes together. I also like to think of learning to play a new chord or riff on the guitar. It begins as a complex set of individual movements. After practice, however, there comes a point where it all begins to flow together as a single motion. That is the feeling we want.

FUN: A time of playfulness; an experience where you were just having fun. That simple. It doesn't have to be the best time of your life, just a moment of enjoyment (not to mention legal and positive).

YES: Something that you can do competently, reliably and repeatedly; something that you know that you can do well without a doubt. I like to think of tying my shoes. As a child, I had some trouble learning, but now I do it automatically without thinking. Think of something you do well and do easily; especially if it took some effort to learn. Be sure to add an appreciation of how well you do it now, compared to how hard it seemed at first.

Remember that none of these is expected to be the very best time that they ever had or the most important decision ever made, just a good example. The names are labels for these states. Please remind participants not to use their own interpretations of the words to define the states. The descriptive paragraphs are the important definition.

In the first half of the exercise we emphasize choosing the appropriate exemplar using the following criteria:

1) It is a specific moment in time (not a series of times) that may be experienced as a short movie or still picture, ending at the most intense part of the experience.
2) The memory chosen must be an active (i.e., non-passive; what I was doing, not what was done to me.) example of the state.
3) The memory should be emotionally clean. It should not (intrinsically) carry the emotional baggage of regret or bad circumstances.
4) The memory or circumstance should be stable over time and not subject to transformation (like focusing on a present job or relationship that could be lost or destroyed).
5) The memory should be free of substance abuse, illegality or questionable motives. It should be as innocent as possible.
6) All exemplars should be experienced for themselves, without regrets or negative baggage. If a state cannot be used without self-pity or remorse, another state must be used. If a participant

refuses to access states without making such attachments, they should be invited to work one-on-one at a later time.

The single event is emphasized to ensure that an appropriate affect is accessed. By returning to a specific instance, characterized by a specific feeling, we can lay a foundation for a specific feeling tone. Later, if other memories are to be added, they are more likely to be linked to the original exemplar by feeling tone as opposed to visual or semantic criteria.

Most of the remaining criteria are stated to ensure that, as we amplify the felt experience of the state, the participant is not thrown into a negative feeling state. Remorse, regrets, feelings of hurt are not useful for our purposes. Further if a memory is rooted in an ongoing enterprise, the failure of that enterprise may taint the exemplar.

One participant used his decision to take his current job as an exemplar for "Solid." When he lost the job, his responses became unstable. Ideally, if the participant enhances the states to the point where they are content free, this will not be a problem. Because, however, responses are often uneven, it remains an important rule. Self-motivated and well engaged participants should not find this a problem.

For Solid, it is best to find a memory that includes a genuine choice process. That is, it represents a specific instance of making a conscious choice from among several possibilities. The exemplar should not be an emotional response to a highly charged event ("I had to stab him, just a little."). It should not be the response demanded by the emotional content of the situation ("I had to marry her, it was the right thing to do."). We are looking for the felt sense of choice that follows a rational procedure. By anchoring this we will create a standard for right decisions, a basis for comparison in demand situations.

We seek active events for the simple reason that we are looking to awaken choice. Passive feelings may be very powerful but as choice is central to the entire process, we ask them to find active states rather than passive.

Objections are often raised as follows:
1)   Nothing good ever happens to me.
2)   I have no positive memories.
3)   I never have fun.
4)   I've never made a good decision.

More often than not these are either expressions of resistance or lack of imagination. They may reasonably express state-dependent memory problems that make access to positive memories somewhat more difficult. Emphasize that they do not need to find the best memory in the world, just pretty good. If you laughed last week, you had fun. If you drove in heavy traffic, performed a difficult task, engaged in martial arts, watched a movie or had good sex, you were focused. If you got up this morning and did what you must to stay alive and free, you made a good decision. If you have ever learned anything, ever; especially like riding a bicycle or driving a stick shift, you have had that moment of discovery. If you can tie your shoes, write your signature, walk or drive a car, you have an example for Yes.

Suggest that childhood memories of innocent experiences are just fine. Remind participants just to go for the memory in isolation. For all examples, just get into the memory for the sake of how it felt then. Each memory is to be experienced as a discrete moment in time.

For participants who have persisting problems finding exemplars for one or more states, have them access a state that they have already enjoyed, enhance it and from that state "imagine" that they can "open a window" into their memory of a good example for the other state.

Participants may be polled, one at a time, for their examples. Good examples should be complimented. Less good examples should be discussed and refined in a gentle and respectful manner. If someone does not have an example, work with them for a few minutes. If they don't get it, suggest that they listen to some of the other examples and return to that participant later in the session. As one of the aims is to choose one example and use it consistently during this exercise, make sure that everyone writes their example on the worksheet.

In the second half of the exercise, participants are asked to close their eyes and re-access the memories, one by one. For each, they should be gently told to notice if they are in the picture or watching themselves from outside of the picture. Suggest that they step all the way into the picture and notice something new about the memory. Suggest further that they pay attention to the order of things that they experience as the memory becomes more powerfully present. The script from the previous exercise is used to lead them into a deeper experience. It is reproduced here.

It is also important to emphasize that not every experience will have the impact of a photographic memory. Initial experiences are often weak and must be enhanced. At the outset, whatever sense of the memory is available will work quite well.

For the most part, the process that follows is identical to exercise one. Have the participants close their eyes and experience the memory. Let them note just how they get to the memory: what do they notice first, a picture, a smell, a feeling? What comes next and next and next?

At each stage of the experience–for every state–encourage the participant to "Gently" turn his attention towards the feeling. How does it move? What temperature is it? Does it have texture? How does it change, two dimensionally, three-dimensionally, more? What color is the feeling? By repeating these and similar suggestions, the facilitator will gently push the participants into a state where there is only room in working memory for feeling data and the sensory detail of the memory will begin to drop out of consciousness. Encourage the participants to "gently turn" more and more of their attention to the feeling so that the pictures and the sounds and the room drift away. This can happen naturally and without effort on their part.

Once again, it is important to emphasize that we are enhancing the memory to arrive at a feeling. The feeling will then be separated from the memory into an independent state of the organism that carries the flavor of the original experience but none of the details.

Participants who can go directly to the feeling level are to be congratulated. Remember that we are using memories to access feelings. Our aim, however, is to leave the memory and keep the feeling.

Go through the following script for each of the resource states. Take your time with each one and insure that all of the participants have a deep pleasurable experience.

SCRIPT

Notice whether, in your imagination, you are experiencing the memory from within, or watching it from outside like in a movie.

If your memory seems to be just in your head, imagine that you can *step all of the way into it*. As you experience the memory, you may even notice flashes that feel like really being there, focus on these. Take a few minutes to make sure that you are actually in the experience. Once you have the sense of really being there, even if it was only for flashes, come fully back into the present context.

Once you have a sense of what it's like to relive the memory from within, step all the way back into it and get a feel for it. Notice that you can step right into one of those parts where it all came alive. Step right into it. Notice what you are seeing and feeling and hearing. Notice the patterns of tension in your muscles. Notice who is there and how you feel emotionally. Take a few minutes to get really familiar with the feel of being there. Enjoy it. Come fully back into the present.

<u>From this point forward drop the use of the word memory and begin to refer to</u> *the feeling, the experience or the resource.*

Once the client has stepped into the experience, they can then begin to vary the submodality structure of the memory. Instruct them to make the changes in a way that makes the experience work best for them. Let them experiment with each dimension to find a level that feels best.

Go through these one at a time, pause after each to allow for their processing time and, in the early exercises, ask them to describe how each change affects the experience. Tell them to remember the ones that work best.

After each change, ask them to note the change in their felt experience. Each instruction is designed to provide a felt change in the experience and to provide practice in the manipulation of feeling by changing the submodality qualities of the experience. Remind them to take note of the kinds of perceptual changes that make the most positive difference in the experience.

Once more, step all the way back into the experience and get a feel for it. Notice that you can step right into one of those parts where it all came alive. Step right into it. Imagine that you can bring the picture closer and make it bigger and brighter.

Each of you has had the experience of changing the size of a picture on a computer screen and changing the brightness on a television set. Use your imagination, now to make the image closer, bigger and brighter. Notice how that changes your experience. Keep the change that produces the best experience. If the experience lessens put the experience back the way it was.

Notice whether the image is in color or black and white—if it is black and white, turn on the color. If it is already in color, turn up the intensity. Notice how that changes the experience. Find out how much color feels best.

Notice whether the experience is moving or still. If it is still, turn on the motion. If it is moving, turn it into a still image. Notice which one feels best. Keep the change that makes the experience feel best.

Finding Resource States

Notice whether the sound is on or off. If the sound is off, turn it on. Adjust the volume of the sound so that it enhances the experience.

Notice how the sound moves. Notice how the sound moves with the experience. Notice whether the sounds are noises or music or voices. Pay attention to where they come from.

Breathe in through your nose and smell the smells that were present there.

Come back for a moment, shake out the experience and talk to me about what happened. Did that feel good? Did you know that you could do that?

What worked best for you?

> After a few minutes of discussion invite them to just close their eyes and return to the place where they left off and continue as follows.

Now, step all the way back to the point where you just left off. For some of you the memory has gone away and you were just out there floating, that's good, go back there. Go back to the state where you left off and notice how easy this is.

Notice how you breathe in this experience. Notice how you hold your body --the patterns of tension and relaxation that enhance your experience. Adjust your posture, so that it enhances the experience.

Notice how you breathe in this experience and the expression on your face. Adjust your expression so that it enhances your experience.

Return to the experience and zoom right back to the very best part. Turn up the brightness, bring it closer and turn up the volume on the sound. While you do these things, note the path of the energy through your body. As you notice the feeling getting stronger, loop the feeling back through the starting point so that it doubles up as it moves through you. Notice that it moves further, faster and more powerfully.

Pay attention to how that works. Find out how many ways you can enjoy it.

> (Provide a few minutes for self-exploration)

And now come all of the way back.

> Now have them return to the state and quickly enter into the feeling state. As they note the rush of onset (call it a "rush") as the experience reasserts itself, have them draw the energy back to the starting point so that the experience feeds forward through the cycle, increasing in intensity. In the exercise we use the phrase "using imaginary hands."

The following language may be useful:

[LOOP}

Once again, step right back into the place where you left off and feel the rush of feeling as you step back in.

Imagine that you can reach out with imaginary hands and take hold of the best part of the feeling as it spreads through your body.

Take hold of it and bring it back to the place where it started. Push it back through the center so that it doubles.

Continue to push it out through your body and notice how it grows stronger. Grab the best part again and push it back through the center.

Repeat this cycling, faster and faster until the state becomes surprisingly powerful.

- Remind participants to attend to the cycling of the feelings not the picture. It can also be useful to think of stirring or turning the felt sense.

- Remind the participants to focus more and more on the qualities of the felt state.

Overload short term memory with impossible dimensions of feeling: location, texture, spread, depth, breadth, height, temperature, imagined color and imagined sound. As the participants focus on more and more of these, the context and content will be crowded out of working memory and they will be left in a powerful, peaceful ecstasy that carries the flavor and physical tone of the original state. It is a generalized state of autonomic arousal that is framed by the original state.

…And as you turn your attention …just gently turn your attention, … to the center of the feeling, you can begin to notice, … really notice… its temperature, … its color…. Notice whether it makes a sound , … or a hum. And you can notice, really notice,… how the feeling moves…. Whether it is centered in your body, or beyond your body…. Whether it moves in a circle … or a loop … or a spiral…..whether it turns clockwise or counterclockwise … and whether it turns like a wheel …or like a turntable…. And as you notice the pattern of this movement, … you can reach out with imaginary hands … and begin to trace this movement… with those imaginary hands, … and if the movement of the feeling …is not a complete movement, … you can take those imaginary hands … and guide that feeling … through its own pattern, … back into its own center, … so that it grows …. and increases … and flows and multiplies. … And you can use those imaginary hands … to take hold of the feeling …. and move it faster … and faster … through its own center so that it doubles … and doubles again, … and grows stronger … and stronger, … and the pictures fade, … and the memories fade … and you find yourself floating … and resting, … down, … all … the … way … down, …into pleasant, …. safe and ….warm. …Resting ….into your own ability … to feel …. good … now….

Allow participants to remain in state for a while. They may safely be allowed to remain in this state for extended periods.
Gently call the participants back to the present time and place.

Come on back. Reorient to the room and the present context in a way that is comfortable and that allows you to retain the lessons of this exercise in the present context. Come on back. NOW.

An important part of the exercise is the abstraction of the feeling from the memory.
- We begin with a remembered experience to gain access to a feeling state.
- We enhance the memory to increase the felt sense of the experience.
- We then focus more and more on the feeling in order to lose it from the memory and discover the feeling as something associated with the participant's own capacity to feel; independent of external influences.

After a brief discussion, repeat the sequence Beginning at the section labeled [LOOP], adding the words

Step back into the state, zoom right back into the very best part and discover how much you can enjoy that state and how many dimensions of wonder you can find inside."

Sequence Note:

This exercise often stretches over two sessions. In the second session it can be useful to review several of the states and anchor them using instructions from the next exercise. This can keep interest going and provide a dramatic illustration of the power of the techniques.

Behavioral Standards

For this exercise, each participant will be able to name and access a memory exemplar for each of the five states. They will provide a written description of each. They will have named the memory. They will be able to transform the original memories into transcendent, content-free trance experiences. As noted, it is not uncommon for this exercise to stretch over two full sessions.

After practicing each of the resource states at the appropriate level, participants should show clear signs of altered consciousness. Relaxation of facial muscles should be apparent, muscle tonus should be modified. Breathing patterns should be altered. Participants will for the most part fail to respond to loud or disturbing noises. If they began in a poor mood, their mood upon returning to normal styles of consciousness will be more positive. Most participants will take a while to come fully back to normal consciousness.

Exercise 2
Finding Resource States

In this exercise we are going to use past experiences of positive emotions and bodily states to create present resources for focused attention, confidence, decision making, optimal learning, and having fun.

One of the important functions of memory is its capacity to make responses that we have already tried or experienced available in present time. Each memory awakens the experience of a situation and a present time experience of the feeling that accompanied it. Both the experience of the memory and the feeling that it produces are present time events. Memory recall is, in a sense, a kind of present time practice of the skill or experience recalled. Some Buddhist teachers have pointed out that, because of the processing time between sensory input and conscious awareness, even present time experience is already in the past (Nyanaponika, 1993).

Our store of remembered experiences represents a set of resources waiting to be tapped. Within that store we can find answers that can be used to solve many of life's problems. Resources can be thought of as any experience or any memory of an experience that you have had. It might even be an imagined experience or a role play. The idea that people possess these kinds of resources is one of the basic presuppositions of NLP (Andreas & Andreas, 1987; Andreas & Andreas, 1989; Bodenhamer & Hall, 1998; Bandler & Grinder, 1975b, 1979; Bandler, Grinder, Dilts, et al.; Haley, 1973; James & Woodsmall, 1988; Linden, 1997).

We will be using remembered and imagined resources for several purposes.

1. As we did in the last exercise, we will be using memories to access a feeling. In this exercise we will continue to use the brain's own control system to enhance the memory, recreate the felt emotion and then experience the feeling in a pure state.

2. We will begin by finding a set of five states that will serve as a set of feeling-tools in everyday life and as a foundation for other exercises.

3. An important purpose of this lesson also continues from the previous exercise: we are learning to make feelings and emotions something that we can do on purpose – not things that just happen to us.

4. In every exercise, learning how to feel good is an important goal. We have all had far too much practice focusing on the other extreme.

Start by finding one memory example for each of the following categories. As you think of the example, pick the best few seconds. For each memory, give it a name and write a short description. The states are: Focus, Solid, Good, Fun and Yes. If you feel that the memory is too personal to write down, just write a word or two that will allow you to remember the same episode in a reliable way (Baffa, 1997; Gray, 2001, 2002).

We are not concerned about which example you find first. The order is unimportant. Feel free to try your favorites first. Come back for the others later. We will be using these examples to access feelings. As we enhance the memories, using techniques from the last exercise, the feelings associated with them will become very strong. This makes it very important to follow the instructions carefully.

Because we will be amplifying the feelings associated with them, use these rules for choosing the memories:

1. Each memory should be positive – they make you feel good.
2. Each memory should represent a completed activity – not something that is still subject to change.
3. The memory should not be related to drugs, illegal activity or sad experiences – even if the decision was otherwise very good and very important.
4. It is OK to choose simple, unimpressive memories. Remember that we are going to take the feelings and magnify them.
5. Make sure to start with one specific example from a single time and a single place – even it was something that you did often. Having a specific memory is more important than having a perfect one.

These are the basic states:

<u>**FOCUS**</u>: This should be an example of focused attention. It might have been a time when you were watching an exciting movie or reading an interesting book. It might be a time when you were playing a game or doing something exciting. Whatever it was, it was a time when you were really THERE, time disappeared and you enjoyed it totally. It might be a time when you met someone new that you were very interested in getting to know. A time when you were able to spend hours with that person, but it only seemed like minutes. Make sure that it was something that you enjoyed. Choose a memory with no regrets or mixed feelings.

Take a moment and write your example down.

<u>**SOLID**</u>: A time when you made a good decision, one that continues to be satisfying even today. Maybe you bought a car or a house and it proved to be a good investment. For our purposes, find an example that involved a real choosing process. Find one that began with many possibilities but ended with a single choice. For example, think of going to buy clothes. Think of the initial choice of a store; then, of a brand or price range; then, of a style. At some point, the choice narrows to just a few possibilities. As you make the final decision there is a feeling that tells you, "This is it." That is the feeling we want.

Take a moment and write your example down.

<u>**GOOD**</u>: A time when you totally surprised yourself by being able to do something, and do it well, despite the fact that you didn't think that you would be able to do it. What might make this experience special is that it wasn't until you had already learned and were already doing the new thing that you finally realized you were doing it! I often think of trying to learn to ride a bicycle or drive a stick shift. There is that one minute when it all comes together. Find a moment like that, when something difficult suddenly

comes together. I also like to think of learning a new chord or riff on the guitar. It begins as a complex set of individual movements. After practice, however, there comes a point where it all begins to flow together as a single motion. That is the feeling we want.

Take a moment and write your example down.

FUN: A time of playfulness; an experience where you were just having fun. That simple. It doesn't have to be the best time of your life, just a moment of enjoyment (not to mention legal and positive).

Take a moment and write your example down.

YES: Something that you can do competently, reliably and repeatedly. It is something that you know that you can do well without a doubt. I like to think of tying my shoes. I had some trouble learning how but now I do it automatically without thinking. Think of something you do well and do easily; especially if it took some effort to learn. Be sure to add an appreciation of how well you do it now, compared to how hard it seemed at first.

Take a moment and write your example down.

Do the following for each example

Go back to the memory and step all of the way into it. Close your eyes and re-experience the whole thing. Provide the memory with a short descriptive name. Write it in the description box above.

Go back to the memory and notice how quickly it comes on. Notice how much more detail is available this time. Notice something that you hadn't noticed before. What are you noticing now that makes the experience more intense? Enjoy your ability to enhance your own experience.

Use the methods from the last exercise to make the memory much more enjoyable. As you step into the memory, rush right to the best part. As you do, make the picture huge and bright and bring it close. Turn up the volume on the sound and notice where the sounds are coming from. As all of this is going on, notice the feelings that you have in your body. Use imaginary hands to take hold of the feelings and double them back through their source. Spin them faster and faster until the feeling becomes very intense. Spin the feeling faster and notice how the details of the memory increase just before they fade and you find yourself floating in pure feeling.

Further Applications

Grow a capacity for focused attention. Become aware of some object, thought or memory. Focus on it and the feelings that it arouses in you. Notice its color and size and the sounds it makes or might make. If you get distracted, practice noticing the distraction—allow yourself to be amused— and gently turn your attention back to the object. Notice how natural distraction is and how quickly it goes away when you just accept it. As you return your attention to the object, notice that it has many features with many dimensions. As you return to it again and again, find more and more ways to experience it. Before long you can become fully absorbed. With a little practice you can turn this ability anywhere.

Create exotic states from simple memories. Use other memories and apply these techniques to them. Discover what happens when small pleasant feelings are magnified. What happens when you find really good feelings and magnify them further?

Discover how much choice you have in guiding your feelings. Build a repertoire of practiced good feelings. Notice how they tend to attract similar memories. Aspiring actors might practice spinning up a series of emotions at will.

References for the exercise

Andreas, C., & Andreas, S. (1989). *Heart of the mind.* Moab, UT: Real People Press.

Andreas, S., & Andreas, C. (1987). *Change your mind, — and keep the change.* Moab, UT: Real People Press.

Baffa, Carmine. (1997). *IQ, hypnosis and genius.* http://Carmine.net/geni/geni0001.htm

Bandler, R., & Grinder, J. (1975b). *Patterns in the hypnotic techniques of Milton H. Erickson, MD, Volume 1.* Cupertino, CA: Meta Publications.

Bandler, R., & Grinder, J. (1979). *Frogs into princes.* Moab, UT: Real People Press.

Bodenhammer, B. G., & Hall, L. M. (1998). *The user's manual for the brain: The complete manual for neuro-linguistic programming practitioner certification.* Institute of Neuro Semantics.

Gray, R. M. (2001). Addictions and the Self: A self-enhancement model for drug treatment in the criminal justice system. *The Journal of Social Work Practice in the Addictions.* 2(1).

Gray, R. M. (2002). The Brooklyn Program: Innovative approaches to substance abuse treatment. *Federal Probation Quarterly, 66(3),* December 2002.

Haley, J. (1973). *Uncommon therapy.* NY: W. W. Norton.

James, T., & Woodsmall, W. (1988). *Timeline therapy and the basis of personality.* Cupertino, CA.: Meta Publications.

Linden, A., & Perutz, K.(1998). *Mindworks: NLP tools for building a better life.* NY: Berkley Publishing Group.

Nyanaponika Thera and Bhikku Bodhi. (1993) *Abhidhamma Studies : Buddhist explorations of consciousness and time.* Somerville, MA: Wisdom Publications.

Anchoring Resource States

Presuppositions Underlying the Exercise

One of our essential goals is to restore a sense of control, choice and mastery in the behavioral options of each participant. In doing so, we seek to open options for new behaviors that are not focused in problematic patterns. In order to produce this sense of mastery, we have chosen to use simple conditioning techniques to teach a level of choice that most people are unaware of. Briefly, each participant is taught how to access several very specific bodily states or moods and to connect those responses to conditioned stimuli with which they can evoke them at will. This starting point has several benefits:

1) It emphasizes the general health of the individual system,
2) It provides the individual with access to powerful, positive, non-problem states,
3) It restores the tendency to see and reexperience positive resource states as a natural part of life, and
4) It provides an opportunity to establish a yes-set in which the exercises are experienced as positive, impactful and entertaining.

More striking than these, the technique provides a specific experience of the ability to control feelings and urges. For any person with compulsive problems, the firsthand experience of emotional control is a milestone in personal growth, self-esteem and the perception of efficacy.

This exercise makes use of a standard NLP technique called Anchoring. Anchoring is a variant of simple associative or Pavlovian conditioning. It uses five examples of general autonomic arousal derived from imagined stimuli as the unconditioned stimuli. In this exercise, we begin with the foundation states already practiced in the preceding exercises. However, if you are using these exercises outside of the Program context, feel free to use any positive experience for the target state (Andreas & Andreas, 1989; Andreas, & Andreas, 1987; Bandler, 1985, 1993; Bandler & MacDonald, 1987; Bodenhammer & Hall, 1998; Brookes, 1995; Dilts, 1993; Dilts, Delozier, Bandler, & Grinder, 1980; Dilts, Delozier, & Delozier, 2000; Gray, 2001; Grinder & Bandler, 1975; Linden & Perutz, 1997).

One of the important things about a positive set of exercises like these is their innate capacity to destroy resistance on multiple levels. First, the exercises are pleasurable. In a group setting, everyone can see the positive responses of the other participants. Second, because the content is private, there is no need for embarrassing self-revelation. Third, by their very nature, the conditioning exercises attach to the leader a strong positive set of associations. S/he becomes a conditioned stimulus for all of the states.

Fractionation is a hypnotic deepening technique in which the hypnotist brings the subject back up towards normal waking consciousness on several occasions during the trance induction. This is done in order to affect a deeper trance experience (Bandler & Grinder, 1975; Hammond, 1990). Practically speaking, the deep regression to the remembered state and subsequent return to the 'here and now' provides a basis for comparison between the states. This allows for better discernment of the qualities of the desired state and enhances the conditioning experience. Functionally, it is identical to fractionation.

The specific aim of this exercise is to provide an experience of efficacy that directly contradicts the feeling of powerlessness that compulsive behavior creates and that much of addictions literature perpetuates. It is important, however, that the icon of hopelessness be approached obliquely. We first establish hope through personal experience, revivification of past positive experience and repeated practice of success. Just as substance abuse and addiction spread through the normal experience of generalization, we use generalization to reclaim psychic ground from them without reification of the process of substance abuse or addiction.

By the end of the session, each participant has experienced the conditioned evocation of several of the target states. More importantly, we have also trained the participants to perform the exercise. In the course of the session many participants will notice increased speed in their own ability to create the anchors. We assign the remaining states for practice at home. This ensures that the capacity to control one's own feeling states is not just connected to the treatment context but will generalize to other contexts. At home, practicing the conditioning exercise, the clients discover that they can do this for themselves and that it has nothing to do with being in that room or with that therapist or anyone else. It is something that is uniquely theirs.

On a deeper level, the exercise provides the next level in a growing sense of self-efficacy; in this case, the sense of the capacity to control their own feelings. We don't have to tell the participants that the next time an urge comes they have a choice. They are being taught not only that they can choose, but how to have choice. On an implicit behavioral level this is emphasized and reemphasized in the first several sessions. Every participant is led to the point where they can experience the five states reliably and repeatedly (Bandler & Grinder, Erickson & Rossi, 1980; 1975; Gray, 2001, 2002).

In some cases we have had clients return and request a state that is specifically designed to overcome the urge to use. With their cooperation, the state is designed and installed. In such cases we ask the following kinds of questions: How would you have to feel in order for that behavior to become unimportant? When have you had an experience of feeling so good or being so fully engaged in something else that the target behavior was not an issue? If there is no single resource state that meets the bill, we can create one from component states. In these cases, we recommended extended practice of creating and anchoring resources, before they are used to counter addictive urges. It is important that the resource be over-learned and valued in its own right before being used instrumentally.

In the course of installing these states, it is important that each experience foster as near a pure ecstasy as possible. Every participant needs to be challenged to discover just how good they can feel. The behavioral logic is simple. If we want sober people we must give people a positive experience of sobriety. How much can you enjoy focused attention? If it were focused in a sexual act, what would be good about it? How good a decision can you make? How much satisfaction can it provide you? How good can you feel about understanding something that you nearly gave up on? How can you really enjoy your capacity to learn?

When the states are sufficiently intense, they lose their association with the original memories and become relatively undifferentiated ecstasies, characterized only by the root frame of the initial state (focused attention, decision making, etc.). At this level they reinforce sober living by providing (simple Skinnerian) reinforcement, as well as the generalizing experience of positive sobriety. Moreover, at deep levels, the states provide a new level of positive experience, a basis for comparison against which the value of destructive behavior can be gauged. We understand this change as a reactivation of the attentional functions of the dorso-lateral prefrontal cortex and the evaluative and self regulatory functions of the anterior cingulated cortex, ventro-medial frontal cortex and the orbito-frontal cortex. More generally we believe that these exercises reawaken the evaluative mechanisms and the default mode network more generally. We also rely on Bandura's observation that sufficiently empowering efficacy experiences tend to generalize to other contexts (Bandura, 1996; Bechara, et al., 1999, 2000; Brown, Manuck, Flory, & Hariri, 2006; Damasio, 1994, 1999; Gu et al., 2010; Li & Sinha, 2008).

Expected Outcome

Participants will learn to install a conditioned response using kinesthetic anchors. They will learn how to enhance the positive aspects of the experience and evoke them at any time. In the process, participants will begin to experience a growing sense of empowerment and choice.

Instructional Notes

By this time, the participants have practiced each of the five states. They have also learned the basic tools for enhancing them. Insofar as they have practiced the techniques, many will be able to directly access the feeling state, independent of the memory content. Because of practice effects, all participants should have the ability to access a strong experience of the identified memory and have experienced some increase in detail and speed of access. Many will experience the state as content free.

There are three crucial issues in the execution of this exercise:

1) Emphasize that your instructions *must* be followed. We have chosen meaningless gestures that are easy to memorize. They also provide a clear signal to the facilitator as to the activities of each participant. Remind them that they will be checked for the use of the appropriate gestures in the one-on-one sessions.

2) It is crucial that the participants do not understand the exercise as the structuring of a simple mnemonic. From time to time people have mistakenly thought that the gestures were intended to act as simple reminders, like tying a string around your finger in order to remember the underlying memory. Nothing could be farther from the fundamental idea of the exercise. The gestures are to be associated with the feelings as conditioned stimuli, triggers for the onset of the feeling that they represent. The end result is to be an unconscious and automatic response. Explain that the gestures will become buttons that will automatically evoke the feeling, even when they feel stuck in another mood.

3) Timing of the anchor is crucial. The gesture, the neutral stimulus action, must be made as the state is increasing. The fingers should touch as the state is increasing, they should separate before the peak. As an alternative, participants may find that keeping the gesture and "pulsing" it or rubbing it works better.

There are four stages to this exercise; installation, enhancement, separation of the feeling from the memory (for those who do not yet have a content-free experience) and testing. The installation is relatively simple. It works straightforwardly as indicated in the exercise itself.

For each of the trials, have the participants begin with their backs straight, their feet flat on the floor and their hands unfolded. Remind them further that erect posture will allow them to become aware of the tendency of each state to suggest a posture that will enhance their experience. As the state increases, they should go with the posture it suggests. Remind them that they should not be eating.

For the first trial, have the participants access the state but remind them to make no effort to make a gesture. Watch their hands and expressions. As they enter the state, remind the group to use no gesture this time; just immerse themselves in the experience. Do it quickly, note the rush of the onset of the experience.

Take them in and bring them back. If they are still working at the level of memory content use suggestions like: "Step all the way into the experience, see what you saw, hear what you heard, feel what you felt." "Take a deep breath and smell what you smelled there." "Notice the rush as you zoom into the very best part of the experience."

For participants who are no longer engaged with content, suggest that they notice how the feeling spreads and grows. Encourage them to notice how deep it goes and how wide. Suggest that they pay attention to multiple dimensions of the felt experiences. Synesthesias can be very useful – If the feeling had a color, what would it be? What kind of sound does it make as it gets stronger? Encourage them to gently turn their attention to the best part of the feeling.

When they have had sufficient time to fully enter the state (note their physiological responses), have them return physically to a neutral state. Have them actually shake their bodies or think of something that will break state. It is crucial that they return fully to a neutral mood between iterations. This fractionation enhances the contrast between the target state and the neutral state and makes the process more effective.

For this exercise we have provided specific gestures for each state. The gestures are meaningless but uniform for the practical reasons stated above. If you decide not to use these gestures or are using the exercise in another context, settle on gestures before doing the exercise.

Before the second iteration, review the gesture. Have them practice making the gesture a few times before doing the rest of the exercise. When they start the next part of the exercise, remind them not to make the gesture until they experience the rush of the state as it comes on. For those in whom the experience starts in the head, let them wait till they begin to feel it in their body before making the gesture. Participants should then hold the gesture for no more than two seconds and then release it. Remind them that they should only hold the gesture while the state is increasing. After releasing the gesture, they should enjoy the state for few minutes before shaking it out. Watch the participants' hands and physiology as they perform the exercise. This will help you to know when to call everyone back to neutral ground.

The anchor gestures (conditioned stimuli) should be made, gently, casually and comfortably. People who are holding them in the air or who seem to be using brute strength must be reminded to relax and to focus on the feeling, not the gesture.

Be sure to emphasize the sequence:

1. Access the state,
2. Make the gesture,
3. Hold the gesture for only two seconds (as it is increasing),
4. Release the gesture,
5. Enjoy the state,
6. Shake it out,
7. Open the eyes.

After each iteration, poll the group for responses and take the time to clarify issues, commend exceptional responses and respectfully correct errors. Remind the participants that the gesture should be easy, casual and should require little attention. Encourage them to focus most of their attention on the state itself. Also make sure to remind them the first few times that 'nothing should happen yet'. They are learning a skill that requires a little practice. You may also want to warn them that the first few times, the gesture will seem to interfere. Encourage them to continue until it enhances the state.

Send them back into state reminding them that they will know that the anchor is working when they note some facet of the experience intensifying as they make the anchor gesture. For each repetition, remind them that they can step directly into the level of intensity at which they last experienced the state. Suggest that they become aware of some new aspect of the experience. "What can you notice now, that you hadn't noticed before? Is there some new detail of feeling?"

For the first anchoring exercise, you may expect five or more trials before the majority of participants report that the anchoring gesture has made a noticeable impact on the experience. After this happens, beginning with the next repetition, instruct the participants to begin 'pumping' the gesture. That is, when they notice the change in experience that accompanies the gesture, they should quickly break and remake the gesture. Each such 'pump' should be followed immediately by a new rush. Let them again break and remake the gesture. Continuing in this manner, they can now time the rhythm of the Anchor gesture to connect with the rushes that they have generated through conditioning effects. Have them notice the increases as they build one increase on another and 'pump' the gesture in time to those increases. This will result in a marked increase of the power of the experience.

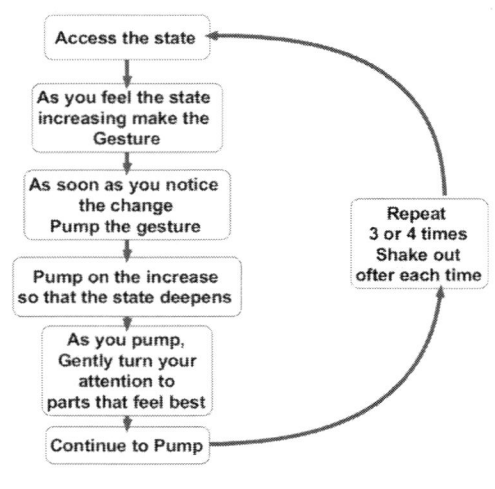

Suggest to the participants that in successive iterations they can time the pumping gesture so that it coincides with the first hint of feeling. Have them turn their attention more and more to the felt bodily sense of the emotion and to pump at the slightest perceived change in the feeling.

The pumping action most often consists of breaking the gesture briefly by separating the fingers. It might be done by tensing the fingers or gently rubbing them together.

There are several comments that may be offered as the intensity increases. Some participants will report that they feel a complete disappearance or significant diminution of the experience between Anchor gestures. This usually means that they are waiting too long between gestures. They should speed up the pumping action so that the release of the gesture is very short. They can also be reminded to make the gesture as the new level is increasing and to quickly release and remake the gesture before it peaks. The idea is that rush builds on rush, which builds on rush, which builds .... It is as if one were riding the crest of a fractal wave.

The physical action of pumping the gesture will not work identically for everyone. Some will prefer a rubbing motion; others will prefer a squeezing or tapping motion. The timing and consistency of the action are the more important elements.

Once the anchor has been associated to the stimulus, there is another refinement that ensures that the anchor is feeling-based and not image-based. This allows the experience to be more transferable to other contexts.

Have the participants sit up straight in a neutral state. Without making any specific effort, have them close their eyes, make the gesture, and notice what happens in their body. Instruct the participants that, as they begin to feel the experience coming up through their bodies (but only after it has arisen spontaneously), they should begin pumping the gesture. As they do so, let them gently turn their attention towards the feeling part of the experience. Suggest to them as they pump, "Just let your attention rest into the very best part of the experience." "Notice how the feeling intensifies, the picture fades and all that is left is the feeling, growing and intensifying." Emphasize the feeling dimensions: depth, texture, color, speed, moisture, etc. Let the participants practice this and remind them that the anchor will intensify the part of the experience where their attention lies. Have them notice that by gently turning their attention towards "the very best part", it will become the center. Have them practice this until all content and context dissipate and only feeling remains. Allow them plenty of time in state to develop depth of feeling.

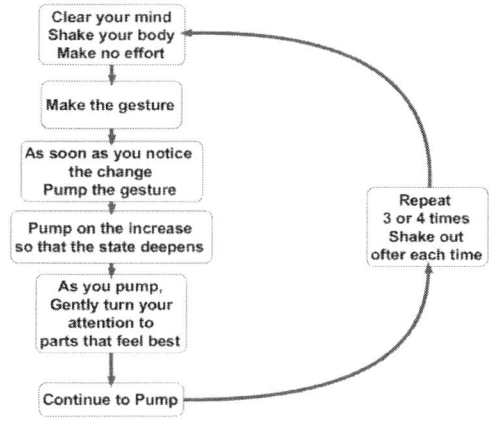

Please emphasize that all of these processes are very gentle and require little effort. Like classical meditation, the attention may be gently turned back to the object when distracted. Further, the anchor gestures should be made comfortably and gently. It is not a test of strength.

By focusing more and more on the feeling aspects, we make use of the limited capacity of short term memory to squeeze out the other parts of the experience. As more and more of consciousness is focused on the feeling, the rest of the experience fades away and the conditioned response becomes feeling centered.

The remaining states (after the first) should move more quickly. Remind the participants that they are learning a skill and that they should notice that it works faster and better as they practice. You might even challenge them to make it work faster or to see how quickly they can get it to work. Review all of the states if possible and assign any unfinished states as homework. In any event, instruct the participants to practice the Anchors at home. Challenge them to see how much intensity they can create.

Behavioral Standards

In the process of conditioning the first five states, there are several things to look for in the participants.
1) Most simply, the participants who are at least paying attention will be able to name the five states and illustrate the appropriate hand gestures. They will be able to do this in order.

2) Physiological signs: Participants who enter the states will show profound changes in posture, heart rate, breathing and skin tone. Many will begin to express rhythmic movements that reflect the underlying experience. Although they differ from person to person, state changes will be observable. As the states become more profound, mouths may hang open and signs of deep relaxation may appear. Smooth, coordinated, consciously regulated movements should largely disappear. Pumping responses may become irregular or intermittent.
3) The anchoring exercises, almost without exception, lead to strong positive feelings. Even in persons who are initially in a bad mood, the exercise is generally sufficient to change the mood. Dissimulators or non-participants typically retain their negative disposition.

By the end of the exercise, participants should be able to change states by pressing their fingers together in any of the five gestures. Persons who have it wrong, who have treated the gestures as mnemonics, will often take a minute to relax and get into the memory. They will also describe the process as going to the memory first in order to get to the feeling. Stop here and re-instruct the group in the proper sequence. Tell them to close their eyes, blank their minds, and without effort, let them just notice what happens in their bodies as they make the gesture. As soon as they notice the slightest change in feeling let them begin to pump the gesture. The only effort should be the relatively passive effort of noticing what feeling arises in the body.

Meditation

The anchoring exercise provides a proper end to the session

Exercise 3
Anchoring Resource States

In the previous exercises we have learned something about the nature of memory and the brain's control system. We have explored the possibility of enhancing memory using that control system and we have learned to access five powerful, altered states. This is all quite impressive, but—all in all— it amounts to a systematization of things that we all do anyway. We all have daydreams and fantasies that can be very intense. We have all spent enough time with them to change our mood. The intensity of the states may be notable and the states may be new, but they are available in other contexts and with other techniques.

We now come to the first element that truly sets this Program apart from the rest. It is simple. It is basic to every living animal in existence. It is something that happens to us every day on an unconscious level. It is conditioning.

Conditioning is one of the most basic forms of learning. It happens to you every day. It is the process that connects a meaningless stimulus to a feeling or response that fills it with meaning. It is how the scent of a certain perfume brings a person to mind, how a song brings back romantic memories and how, after food poisoning, you can never stomach crab cakes again.

All of this happens out of consciousness. Conditioning usually works and works best when we are unaware of it. We usually only recognize it after the fact and call it learning or association. Here, we are going to create our own conditioned responses.

In this exercise we are going to learn how to connect the five states that we have just learned — focus, solid, good, fun and yes – to five simple gestures. We will then be able to use these gestures to evoke their associated feelings at any time and in any place. We will also be able to use the states in much more complex ways because of this accessibility.

This step is essential because changes that happen in meditation or therapy often don't last long outside of the original experience. It is not uncommon for someone who is meditating or getting acupuncture to have wonderful results for hours or even days after the experience. When it is part of a disciplined practice, these experiences can continue for quite a while. The problem, however, always remains: what happens if you can't make the session? What happens when the world smacks you in the face and there is no opportunity to stop for meditation or to see the massage therapist or the acupuncturist? What happens, if you can't assume the posture or otherwise access the state?

Conditioning allows you to take the state with you. It compresses all of the work of getting there and enfolds it into a word, a gesture or a smell. In this context we have chosen to think of them as buttons. This is very much like the common expression for conditioned responses – "pushing someone's buttons". Conditioned responses are like buttons that can turn on the feeling states. In this case we will be turning on positive buttons, buttons that you can use to attain ecstasies of transcendence, buttons that can bring you from negative to neutral, buttons that can give you the edge you need in any context. They are also buttons that you can use to construct states of mind that you may have never experienced before.

Buttons also give you a new means of controlling the intensity or quality of the states. Once established, the buttons can be used to focus the states on facets of the experience that you particularly enjoy.

At the outset, please remember that this is easy. It is a natural process that is best experienced somewhat passively; straining and struggling and conscious efforts only get in the way. During the exercise, just relax and notice how well your body responds.

By this point, you will have practiced several of the states to the point where you find it easy to access empowered, peaceful experiences of floating ecstasies. If you are starting from here, you can go directly to step three.

If you are working on a new state or one that still needs practice, start as follows:

Creating Anchors

It is very important to notice the difference between being *in* the memory and watching the memory from outside. If your memory seems to be just in your head, imagine that you can step all of the way into it. As you experience the memory, you may even notice flashes that feel like really being there, focus on these. Take a few minutes to make sure that you are actually in the experience. Once you have the sense of really being there, even if it was only for flashes, come fully back into the present context.

1. Step all the way into the memory, and get a feel for it. Notice that you can step right into one of those parts where it all came alive. Step right into it. Notice what you are seeing and feeling and hearing. Notice the patterns of tension in your muscles. Notice who is there and how you feel emotionally. Take a few minutes to get really familiar with the feel of being there. Enjoy it. Come fully back into the present. Do this a few times and find out how much you can enjoy it.

2. Return to the memory. As you do, notice that you can zoom right to point where you left off the last time; right to the very most intense part. Notice the rush of feelings and sensations. Enjoy them for a moment and then, return fully to the present.

3. To anchor the state, pick a gesture (use the ones supplied below). Close your eyes and zoom right back to the most intense experience of the state. As you experience rushing into the state, make the gesture. Hold it for about two seconds while the feelings are increasing. Let the anchor go. Shake out the state (shake your body) and return to the present.

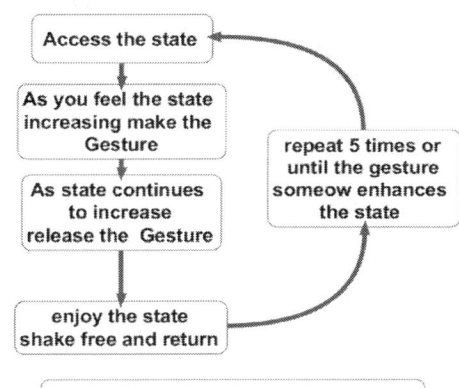

Self-Anchoring Process

(Here is the basic sequence: close your eyes, access the feeling, make the gesture, release the gesture, shake out the feeling, open your eyes.).

4. Repeat step three, four or five times or until you begin to notice a dramatic change in the experience whenever you make the gesture. Remember to come fully back into the present before each repetition.

5. Once you have the clear sense that the gesture is adding to the power or depth of the experience, repeat step three with the following difference: As you notice the change that flows out of making the gesture, quickly break and remake the gesture. Hold the gesture again until you become aware of the rush of experience then break and remake the gesture again. Repeat this pumping action until the experience becomes pleasurably intense. Shake out the state (shake your body) and return to the present.

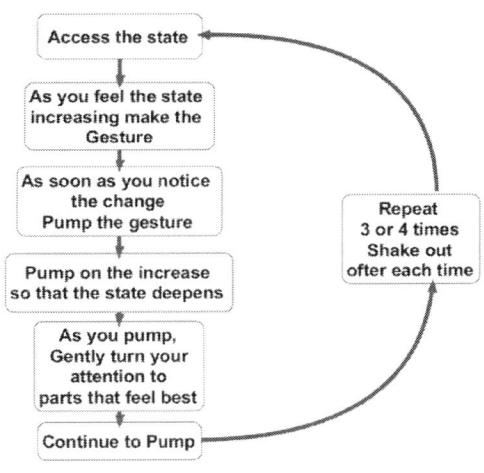

Enhancing the Anchor

TIP: *Pumping the gesture might mean gently rubbing the fingers together; it may mean gently pulsing the muscles while holding the gesture. I generally find that once the anchor has been created, pulsing the gesture works best. Find a method that works for you.*

Notice that as you focus your attention on the very first sensations that enter your body, the feeling grows more quickly. Make a game of pulsing or pumping the gesture just as the first hint of feeling arises in your body.

6. Test the state. Clear your mind. Sit or lie comfortably and make the gesture. As you notice the rush of feeling that flows out of making the gesture, begin to pump the gesture by breaking and remaking the gesture repeatedly. Do your best to make the gesture at the first hint of a bodily feeling. Repeat this pumping action as you find yourself enjoying the growing intensity of feeling. With each

pump, allow your attention to discover something better or deeper in the feelings. As you do this, enjoy more and more aspects of the feeling itself. Let your attention move into the feeling and just allow the visual and auditory information to fade away. Keep pumping until you have an intense experience of pure feeling. Shake out the state (shake your body) and return to the present.

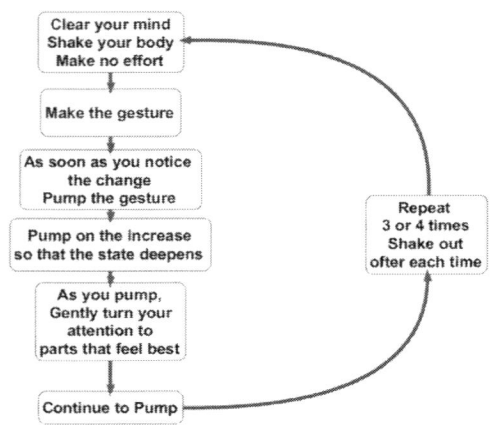

Testing the Anchor

The basic states

FOCUS: This should be an example of focused attention. It might have been a time when you were watching an exciting movie or reading an interesting book. It might be a time when you were playing a game or doing something exciting. Whatever it was, it was a time when you were really THERE; time disappeared and you enjoyed it totally. It might be a time when you met someone new that you were very interested in getting to know. A time when you were able to spend hours with that person, but it only seemed like minutes. Make sure that it was something that you enjoyed. Choose a memory with no regrets or mixed feelings; something that is complete in itself. (Touch tip of thumb to tip of index finger.)

SOLID: A time when you made a good decision, one that continues to be satisfying even today. Maybe you bought a car or a house and it proved to be a good investment. For our purposes, find an example that involved a real choosing process. Find one that began with many possibilities but ended with a single choice. For example, think of going to buy clothes. Think of the initial choice of a store; then, of a brand or price range; then, of a style. At some point the choice narrows to just a few possibilities. As you make the final decision there is a feeling that tells you, "This is it." That is the feeling we want. (Touch tip of thumb to first joint of index finger.)

GOOD: A time when you totally surprised yourself by being able to do something, and do it well, despite the fact that you didn't think that you would be able to do it. What might make this experience special is that it wasn't until you had already learned and were already doing the new thing that you finally realized you were doing it! I often think of trying to learn to ride a bicycle or drive a stick shift. There is that one minute when it all comes together. Find a moment like that, when something difficult suddenly comes together. I also like to think of learning a new chord or riff on a guitar. It begins as a complex set of individual movements. After practice, however, there comes a point where it all begins to flow together as a single motion. That is the feeling we want. (Touch tip of thumb to tip of middle finger.)

FUN: A time of playfulness. This is an experience where you were just having fun; that simple. It doesn't have to be the best time of your life, just a moment of enjoyment (not to mention legal and positive). (Touch tip of thumb to first joint of middle finger.)

YES: Something that you can do competently, reliably and repeatedly; something that you know that you can do well without a doubt. I like to think of tying my shoes. I had some trouble learning how, but now I do it automatically without thinking. Think of something you do well and do easily; especially if it took some effort to learn. Be sure to add an appreciation of how well you do it now, compared to how hard it seemed at first. (Touch tip of thumb to tip of ring finger.)

Further Applications

Anchoring is the basic way in which organisms learn. There is virtually no end to the ways you can use it. It is a valuable tool for moving deeply felt experiences from one context to another. Once you have mastered the technique using the five basic examples found here, there are all kinds of things that you can anchor: clarity, creative inspiration, ecstatic experiences, and new ways of perceiving the world. Several participants have anchored the feeling of love so that they could expand the experience to all of creation. Anchoring can be used to enhance an imagined feeling, expand a weak one, or to create an experience of how you think someone must feel. In a later exercise we will ask you to create several anchors of your own choosing.

References for the exercise

Andreas, S. & Andreas, C. (1987). *Change your mind— and keep the change.* Moab, UT: Real People Press.

Baffa, C. (1997). *IQ, hypnosis and genius.* http://Carmine.net/geni/geni0001.htm

Dilts, R., Delozier, J. A., & Delozier, J. (2000). *Encyclopedia of systemic neuro-linguistic programming and nlp new coding.* Scotts Valley, CA: NLP University Press.

Gray, R. M. (2001). Addictions and the Self: A self-enhancement model for drug treatment in the criminal justice system. *The Journal of Social Work Practice in the Addictions, 2(1).*

Gray, R. M. (2002). The Brooklyn Program: Innovative approaches to substance abuse treatment. *Federal Probation Quarterly.* 66(3), December 2002.

Exercise 4

Keys to Enhancing Subjective Experience

Presuppositions Underlying the Exercise:

This exercise presupposes that subjective experience can be controlled and manipulated using submodality distinctions and other techniques. It further assumes that participants have anchored the states and have spent some time adjusting the submodalities already. This exercise outlines specific enhancements and systematizes them for the use of the participants.

Once again it is important to note that more control is returned to the participant. They are now given specific tools for the enhancement of the states that they have just automated.

A major presupposition of the exercise is that people want to feel good and that good feelings, even intense good feelings, are not problematic when they are accompanied by a sense of personal control.

We further assume that ecstasy is one of the states that humans normally seek.

In general, this exercise may or may not be distributed to the participants. It is essentially a practice session in which new enhancements for the states are taught to the participants or they are just encouraged to practice the anchored states. It is also a time when the states are deepened and refined. If all of the participants have anchored content free anchors, use the session to review the anchors answer questions and allow the participants to spend as much time as possible using the anchors to explore inner space.

Expected Outcome

Participants will enjoy and be able to implement one or more techniques for enhancing their own subjective experience. They will create increased levels of control and self-efficacy. They will enjoy extending their ability to experience autogenic states.

An important result here is to ensure that the states have been enhanced to the point that each is completely content free. It is crucial for the later exercises for each participant to have anchored relatively pure resource states that no longer have any association, beyond the feeling itself, to the original memories themselves. This will ensure that the resources will blend flawlessly and that they can be used cross contextually without problems. To this end, each of the enhancements beyond the basic state elicitation works to ensure that any trace of content or context is removed fully from the root resource memory.

Instructional Notes

The current exercise, Keys to Enhancing Subjective Experience, begins with a recapitulation of the techniques of anchoring and State Enhancement. The first techniques represent a review of the fundamentals of anchoring and memory access.

Sequence Note: Many of these techniques can be used to enhance the first two exercises and to ensure that before the states are anchored they have been brought to optimal levels. This exercise and the one following can usually be completed in one two-hour session. Nevertheless, using one or more full sessions to access the states and enjoy them is highly recommended.

Behavioral Standards

Participants will be able to use the nine techniques to create and enhance anchored states. While using preferred modes, they will exhibit appropriate affective and physiological changes. Participants will be able at this point to describe how the technique impacts their experience of the states and their progress towards peak.

Meditation

In this session, we end with and extended experience of the anchored states.

Exercise 4

Keys to Enhancing Subjective Experiences

This exercise is mostly review. In the first section, Initial Strategies, the exercise reviews some of the basic techniques of submodality manipulation — the brain's rules for perception. You can use it as a good source of review for creating new anchors.

There are two really new methods for enhancing the felt experiences. Both require you to use your imagination. In an important sense we can think about imagination as the manipulation of our internal experiences. All of our exercises rely on imagination. What is important about this is the understanding that imagination is crucial to the proper function of the mind. Antonio Damasio refers to imagination as "as if circuits." They reflect activity in the places where the brain acts "as if the body was responding" (Bechara, 2000; Damasio, et al. 2000).

Active imagination is an important tool in science. Einstein described his work as mostly done with thought experiments that he imagined and the feelings that they produced. Leonardo DaVinci described many techniques for visualizing the impossible. Creativity and personal growth in all spheres can be linked to the positive use of our capacity to imagine (Dilts, 1995).

Initial Strategies

1) Full sensory evocation. The more senses that you involve in awakening the state, the stronger the state becomes. States are multi-sensory experiences. The more sensory information you can add to the experience, the stronger your experience of the state will be. Keep adding additional information to the experience. If you can see that it's a restaurant, what were the smells? If you can feel the cool breeze, what does a breeze like that sound like? If you can feel the breeze, what would you see in the trees (Bandler & Grinder, 1979, Dilts, et al., 1980, 2000; Linden & Perutz, 1998)

2) Submodalities. Note the submodality differences. Go for things that you don't normally think of like direction, smell (what are its dimensions?), the location and richness of sounds. Pay attention to your physical tension and posture. What is the rhythm of the state? Robert Dilts, one of the founders of NLP, often focuses on movements associated with the state and exaggerates them.

Notice if you are moving in the memory, if you are, express the movement in a subtle manner. Richard Bandler, another founder of the field and the main developer of these techniques, puts a lot of emphasis on distance, location and size. Try making the visual image bigger and closer. For hearing, try noticing the direction from which the sounds come. For feelings, enhance their size, intensity and motion (Andreas & Andreas, 1987; Bandler & MacDonald, 1987; Dilts, et al., 1980, 2000).

3) The difference that makes the difference. Every state has a driving submodality, that is, one sense, or some facet of a sense (brightness, focus, size, motion, etc) that enhances the quality of the entire experience. When you find that one; use it. As you examine each part of the experience, notice one that, when you change it, the whole experience is transformed (Andreas & Andreas, 1987; Bandler & MacDonald, 1987; Linden & Perutz, 1998).

4) Progression and speed. Notice how the experience grows. Where do you first notice it? What sense are you using? Does that change or does another sense take over? What happens next? And then? As you notice the sequence and direction of movement in the sensory experience, discover what happens as you speed up the process. How much more intense would it become if it increased in that same way for two days, in the next ten seconds? What other dimensions does the experience have? What happens if it thickens or spreads as it speeds up (Laski, 1961)?

5) Fractionation. Get into the state. Crank it up. Shake it off and get fully out of it. Go back for it again. Notice the difference. Shake it out. Do it again. Contrast is important. One of the easiest ways to enhance your experience of a state is to get all the way in, and then completely get out of it before going back.

6) Loop through the response sequence. Notice that there is a place where the feeling starts; a specific way in which it grows towards peak, and an area of feeling that marks its greatest intensity. It also has a characteristic way of dissipating or disappearing. When you have a good idea of how the feeling grows and changes, focus on the start of the feeling and follow its spread towards maximum intensity. When you have a good sense of how it moves, re-circulate it through the place where it started. Every time it approaches the place where it peaks, loop it back through so that the energy doubles in intensity. When it again reaches its maximum extent, loop it back through the center. Continue to loop it through your body, faster and faster until it becomes ecstatic (Bandler, 2001).

This technique is easier if you imagine grabbing the energy with imaginary hands and actually pushing it through the starting point and out into your body. As you feel it intensifying, again reach out with those imaginary hands and double it back upon itself. Keep the motion going until you can direct it as a felt sense. Spin it faster and faster.

7) Imagine seeing yourself experiencing the state two or four or many times more intensely. Notice how you look. Step into the image. Swish into the image, that is, have that image suddenly explode into your experience and wash over you. Do it quickly. Carlos Casteneda might suggest that you imagine yourself sitting or lying in the same posture in the imagined state as you actually hold in the present state (Bandler, 1993; Casteneda, 1993).

8) Metaphorical dials and knobs. This really works. Make believe that there is a knob or a clicker switch or whatever works for you. While focused on the very best part of the state, ACTUALLY reach out and turn the knob. Make believe it's there. FRACTIONATE. Turn the knob up, and then turn the knob down a little. Turn the knob up again, and turn it down again. Get a sense of the speed at which your knob works and anchor at the appropriate spot, near one of the peaks (Bandler, 2001).

7) Anchor the onset of the state. Be sure that you anchor the state (make the stimulus touch, gesture, sound, etc.) as the state is beginning. Don't wait until it's already peaked. Anchor the increase of the

desired state before peak, not the peaked or decreasing state. Once you have a clean sense of the state coming on, begin to time the stimulus gesture to anchor and enhance the rush.

When you are first doing the anchoring exercise it is best to set the anchor just as the entire state is moving towards its fullest expression. When you get to the point where the anchor stimulus itself is creating a clear response, begin to pump the anchor. That is: fire off the anchor. Then, as the feeling increases, fire it off again. As one pulse adds to the growth of the next — before anything else can happen, fire of the anchor again. Continue this for a while and notice how strong it becomes (Dilts, et al., 1980; 2000).

As we work to enhance the states, it now becomes important to ensure that you have let go of the memory connection. We used the memory to get to the feeling, now we want to ensure that the feeling available as a pure state, without the memory. Close your eyes and, with no other effort, make the gesture and just notice the feeling that arises in your body. As you notice the feeling; pump the gesture. As you pump the gesture, notice how the feeling grows and expands. Notice the way it spreads. Does the temperature change? Does the feeling have depth or dimension? Where is the very best part of this feeling? Turn your attention to the part you like best and pump it. As you pay more and more attention to the dimensions of feeling, allow the visual and auditory parts of the memory to just fade away. Enjoy drifting in the pure feeling state. As you drift, your brain may generate new scenes or no scenes. Unconscious landscapes may arise, or you may drift in a rainbow mist. You should now find that the states arise with no effort every time you make the gesture.

10) After you've created and practiced an anchor, stop and remember your best experience of the anchor. Notice where the feeling starts and how it spreads and then spin the feeling until it is much stronger than you remember it. At this point, fire off the anchor and begin to pump or pulse as usual. Notice the new levels of intensity.

Further Applications

This exercise is here partly for review. It also suggests some interesting things about perception and memory. Richard Bandler, the source of most of the techniques described, has pioneered creating a series of imaginary levers, dials and knobs to manipulate submodality distinctions. What would happen if you created a control panel for your brain? What if you had a button for the feeling that you get when adrenaline starts to pump or when you're completely satisfied after a good meal, the last cigarette, or great sex?

Can you imagine seeing yourself living out different realities or really sharing someone else's experience? What happens when you literally step into an image of someone else's body? The authors of *Super Learning* describe experiments by a Russian psychologist who had music students use their imagination to step into the bodies of great masters as they played. After imagining how it felt to have that level of skill, the students made significant developmental leaps. Whose talent might you try on?

References for the exercise
Andreas, S. & Andreas, C. (1987). *Change your mind-- and keep the change.* Moab, UT: Real People Press.

Bandler, R. (2001). *Design human engineering.* (Audio).

Bandler, R. (1993). *Time for a change.* Capitola, CA: Meta Publications.

Bandler, R. & Grinder, J. (1979). *Frogs into princes*. Moab, UT: Real People Press.

Bandler, R. & MacDonald, W. (1987). *An insider's guide to submodalities*. Moab, UT. : Real People Press.

Bechara, A., Damasio, H. & Damasio, A. R. (2000). Emotion, decision making and the orbitofrontal cortex. *Cerebral Cortex, 10(3)*, 295-307

Damasio, A. R. (1999). *The feeling of what happens: Body and emotion in the making of consciousness*. NY: Harcourt.

Damasio, A. R., Grabowski, T. J., Bechara, A., Damasio, H., Ponto, L. B., Parcizi, J., & Hichwa, R. J. (2000). Subcortical and cortical brain activity during the feeling of self-generated emotions." *Nature Neuroscience, 3(10)*, October, 2000.

Dilts, R., Delozier, J., Bandler, R., & Grinder, J. (1980). *NLP. Vol.1*. Capitola, CA: Meta Publications.

Dilts, R. (1995). *Strategies of genius (Vol. 3)*. Cupertino CA.: Meta Publications.

Dilts, R. & Delozier, J. (2000). *Encyclopedia of systemic neuro-linguistic programming and nlp new coding*. Scotts Valley, CA: NLP University Press.

Gray, R. M. (2001). Addictions and the Self: A self-enhancement model for drug treatment in the criminal justice system. *The Journal of Social Work Practice in the Addictions, Vol. 2*, No. 1.

Gray, R. M. (2002). The Brooklyn Program: Innovative approaches to substance abuse treatment. *Federal Probation Quarterly, 66(3)*. December 2002.

Laski, M. (1961). *Ecstasy in secular and religious experiences*. NY: Jeremy Tarcher.

Linden, A. & Perutz, K. (1998). *Mindworks: NLP tools for building a better life*. NY: Berkley Publishing Group.

Exercise 5
Getting to NOW

Presuppositions Underlying the Exercise:

One of the major presuppositions of this exercise and the Program as a whole is that the human psyche is a complex system and that system principles give rise to properties that are not necessarily predictable from the individual elements. This exercise illustrates the capacity of unique experiences to constellate a positive sense of Self that is both familiar and new to the individual. It further presupposes that every positive mood or state of mind that becomes consciously available to the individual also becomes a reason to do something other than use illegal or destructive substances (Bertalanffy, 1968; Fidler, 1982; Gray, 1994; Piaget, 1970)

The exercise presupposes that one of the important prerequisites of self-esteem is the possession of a relationship with a deep-enough representation of Self, one that continues over time. On a preliminary level, this exercise awakens the individual to a deeper sense of the choosing Self, the Self that is always there (Gray 1997a, 2001, 2002).

There is also a presupposition that the participants have created the conditioned stimuli for the basic states and that they have learned to enhance those states. Without this foundation, the exercise will not be successful. In each case, the state anchors should all be experienced at a level of pure feeling. Neither content nor context should remain.

The exercise explicitly does not presuppose that the Self is ONLY an artifact of mechanical stimulus response systems. It does, however, presuppose that what we call the Self can be made more conscious and more accessible as a continuing reality through the synergistic interaction of conditioned response systems.

Expected Outcome

Participants will use their growing expertise in creating anchors and following behavioral instructions to assemble a positive resource state as an entity independent of any specific memory precursor. They will be able to enhance the state and anchor it to a specific stimulus motion.

Instructional Notes

Getting to Now proceeds in a straightforward manner from the previous exercises. In order for it to work well, the participants must have successfully created and enhanced the initial five states (Focus, Solid, Good, Fun, Yes). The states should be pure feeling states; they should be fully dissociated from memory associations.

Have the participants review each of the states. Have them fire off each anchor, pump the state up and spend a few minutes enjoying it. Address any problems and make sure that all have mastered the process.

Script for creating NOW

Begin by illustrating the gesture. A gently clenched fist made with the non-dominant hand. Practice the gesture a few times and remind the clients that they already know how to do this as they have already created and practiced other anchors.

Fire off Focus
and allow your attention
to rest down into the very best part.
Gently turn
your attention
into the patterns
of tension and relaxation,
the rhythms
of taking hold and letting go.
And pump it, and pump it and pump it.

Allow your self
to float down and back
into the very best part
and become aware of pattern
of the state,
the tensions and relaxations,
the height and depth,
where it is warm and where it is moist.
How it moves and where
it is very, very still…

Hold that in mind and body
as a single pattern of being
A pattern that you can remember
and rest down into it.

And as you hold that pattern
in mind and body,
move your fingers to fire off Solid.
And continuing
to hold the pattern
of focus in mind,
notice how Solid
arises in your body.
Ad pmp it and pum it and pump it.
Notice how it moves
and weaves
and interacts with Focus.
Notice how its patterns
of tension and relaxation,
taking hold and letting go,
Interweave
and merge
and combine with Focus

Weave them and harmonize them,
integrate them
and notice,
really notice
how they complement each other.
Get a sense of their dance
and the new pattern
that emerges as they
meld and join,
creating
something
new.

Notice how
they compliment each other...
Notice their contrasts. ...
Hold the combined pattern
 in mind and body
as you move your fingers
to the GOOD gesture. ...
Fire it off
and notice how the sensations of
GOOD
dance with
and compliment
the feelings
already present. ...
Become aware
of the changing patterns ...
of tension and relaxation, ...
warmth and light. ...
Notice how they weave
together
into a new pattern.

Weave them and harmonize them,
integrate them and notice, really
notice
how they complement each other.
Get a sense of their dance and the
new pattern that emerges as they
meld and join, creating something
new.

Notice how they complement each
other...
Notice their contrasts. ...
Hold the combined pattern in mind
and body as you move your fingers to
the FUN gesture. ...
Fire it off and notice how the
sensations associated with FUN
dance with and compliment the
feelings already present. ...

Become aware of the changing
patterns ...
of tension and relaxation, ...
warmth and light. ...
Notice how they weave together into
a new pattern.

Fire off YES ...
 and become aware
 of its contribution ...
 to the new patterns of feeling. ...
 Notice how it rises in the midst, ...
 creating a new space, ...
 a new harmony, ...
 a new pattern of feeling. ...
 Become aware
 of its temperatures and patterns, ...
 its silences and sounds. ...
 And, ...
 as you notice,
 really notice ...
 how they all come together ...
 in a new kind of way;
 make a fist, ...
 the anchor for NOW. ...
 Pay attention to the new state
 that is now arising. ...
 gently pump or pulse the gesture. ...
 Use the skill of pumping
  and anchoring
 as you have been
 practicing ...
 Continue to become aware ...
 of new facets ...
 of the experience, ...
 pumping as you go.
 Explore the depth ...
 and height ...
  and breadth ...
 of the new state ...
 and pump it. ...
 Rest into it ...
 and pump it. ...
 Allow yourself to float. ...
 And pump it, ... and pump it, ... and
 pump it, ... and pump it. ...
 Gently ...
 turn your attention ...
 to some wonderful part. ...
 And Pump it into the center.

Notice that you can
gently
turn your attention
to the very best part

> and pump it
> so it expands
> to fill the center.
> Discover something new
> about how much
> deep joy
> can be yours
>
> as you pump.
>
> Spend some time
> enjoying and exploring
> that state
>
> Let the Gesture go
> and come all of the way back.

Repeat the process until the gesture reliably produces an enhancement of the new state. Pump up the state using the same technique described in the previous exercises.

Problems associated with timing or content issues:

A small number of participants have reported dizziness. Dizziness usually results when an anchor is still attached to the visual part of the memory. This suggests that, for one or more of the root states, they have anchored and enhanced the intact memory, not just the kinesthetic elements. Participants may practice accessing the anchor by focusing on the very first hint of felt bodily experience and practice until the anchor is context free. As an alternative, have the participants who report this problem speed up this exercise so that they shift to the next anchor before any visual stimuli appear. That is, after firing off the anchor for the first state, they should move to the next one more quickly– just as they note the feeling arising in their body. In each case they should move at the first hint of a kinesthetic rush and before there is any hint of content.

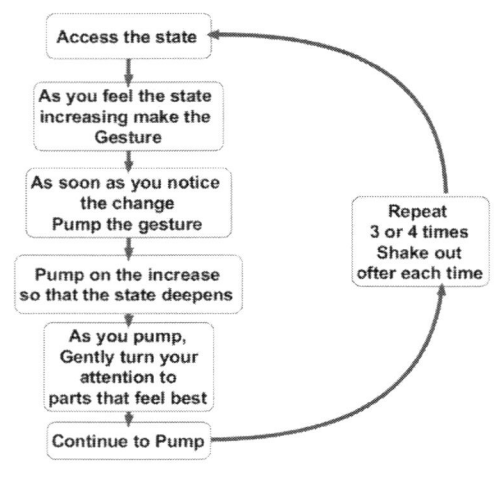

Enhancing the Anchor

Some participants have reported that they are unable to make the transition from state to state, or that the states will not mix. This usually means either that they have come out of one state before going into the other or, again, they have retained incompatible content. Have them speed up the transitions and make sure that they fire off the next state while the one before it is still increasing. It might be useful to tell them to let the states overlap or to keep the feeling of each in mind as they add in the next.

In each such problem evocation, it is wise to repeat the anchoring of the NOW state several times, adjusting the instructions each time to ensure that all of the problems are ironed out. Persistent problems can be handled in a one-on-one session.

In the session when NOW is created, allow the participants to stay in state for as long a time as possible. Encourage them to explore the dimensions of the state and encourage them to discover its depths and heights. Remind them repeatedly to allow the attention to move gently towards a particularly pleasant or interesting facet of the experience and to pump it so that it expands to become the center of the experience. After several repetitions, return and debrief the group.

After debriefing, participants should be instructed to test the state just as they did the others: With no effort, close the eyes, make the gesture, and pump at the first hint of bodily feeling. Continue pumping for 10 or 15 minutes. Recall and debrief.

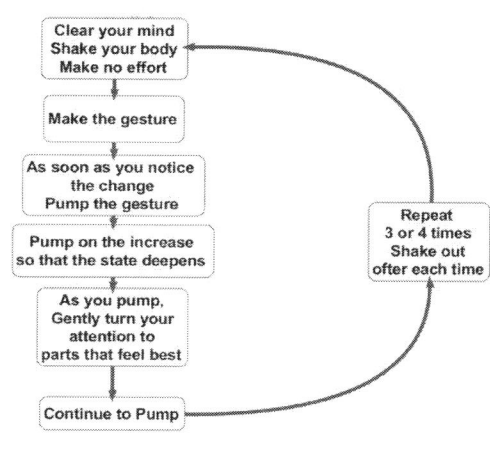

Testing the Anchor

Behavioral Standards

Participants will be able to describe a distinct state independent of the component states. It should be a coherent whole with elements of centeredness, energy and objectivity. It often provides a surprising contrast to the intensity of the component states. Participants will be able to access the state at will and suggest places where it might be useful in their lives.

Meditation

Although we often skip the meditations, here we use the meditation, New Worlds to Gain. This meditation sets up a link between the present experience of NOW and an imagined future experience. It multiplies the depth of the experience using the Ericksonian pseudo-orientation in time discussed in later segments (Erickson, 1954; Gray, 1997b, 2001).

Use the meditation after several iterations of testing and enhancing NOW.

Exercise 5
Getting To Now

The roots of the current exercise lie in the idea that each human life has at its heart a center and a purpose. Discovering that center and the purpose towards which it points are crucial. Carl Jung called the process of coming to know one's self, *individuation*. Abraham Maslow called it *self-actualization*. In general, the first step along this road is the experience of what it means to be you--Just You.

This sounds simple, but in practice it is among the most difficult of tasks. Who we are is often masked by 'shoulds' and 'ought-tos' imposed from without. We have spent years seeking external approval and transient pleasures. It is hidden by the habits of a lifetime spent learning the acceptable ways to act and think as dictated by family friends and associates. More often than not we are not even aware of what is truly important to us as individuals. This exercise represents the first of several that are designed to awaken experiences of your deep self. This first provides a foundation for growth, a starting place for further exploration.

The exercise itself is taken from Jung's idea that our experience is organized and understood in terms of a web of felt associations. The Self is the center and root of that sense (Jung 1979a, 1979b). In more recent neurophysiological research this idea has been vindicated by the work of Damasio, LeDoux and others who point to just such a felt sense and its determination by the activation of specific brain regions (Damasio, 1999; Damasio, Grabowski, et al., 2000; D'Aquili & Newberg, 1998, 2000; LeDoux, 1999, 2000,).

It is also rooted in the idea that every feeling, when felt deeply enough, carries within it the seed of wholeness. Each feeling points down to the true center of who we are. So, we begin with the five resource states that we have been practicing. You may have already noticed that they have awakened a deeper sense of what it means to be you, of what is important and what is not. This is how we experience individuality as it awakens in consciousness. We will now use the five basic anchors, **Focus, Solid, Good, Fun** and **Yes** to create an experience of how these five simple things can create something rather surprising. This is the new anchor, NOW. Its gesture is a gently formed fist. You will use it just as you did the other gestures. Please read through all of the instructions before doing the exercise (Gray, 2001, 2002).

Getting to NOW

1) To begin, elicit the five states, one at a time. Make sure that each one is strong. Make sure that each is a pure feeling state with no remainder of pictures or context from the original memory. Pump them up and shake them out, one at a time.

2) Once you have elicited the states one at a time, begin to stack them as follows:

Fire off FOCUS. Notice how the state arises in your body. Pump it up and pay attention to its feel, where it centers, its patterns of relaxation and tension, its warmth, its character. Hold the felt pattern of these perceptions in mind and move your fingers to the SOLID gesture. Begin to pump and notice how the feelings associated with SOLID combine and interweave with the feelings from FOCUS. Notice how they complement each other. Notice their contrasts. Hold the combined pattern in mind and body as you move your fingers to the GOOD gesture. Fire it off and notice how the specific sensations associated with GOOD dance with and compliment the feelings already present. Become aware of the changing patterns of tension and relaxation, warmth and light. Notice how they weave together into a new pattern. Hold the felt pattern of these perceptions in mind and move your fingers to the FUN gesture. Begin to pump and notice how the feelings associated with FUN combine and interweave with the feelings from the others. Notice how they complement each other. Notice their contrasts. Hold the combined pattern in mind and body as you move your fingers to the YES gesture. Fire it off and notice how the specific sensations associated with YES dance with and compliment the feelings already present. Become aware of the changing patterns of tension and relaxation, warmth and light. Notice how they weave together into a new pattern.

NOW, as you begin to become aware of its contribution to the new patterns of feeling, make a fist, the anchor gesture for NOW. Pay attention to the new state that is now arising. As you do so, gently pump or pulse the fist. Use the skill of pumping and anchoring as you have been practicing it with the other anchors. Continue to become aware of new facets of the experience, pumping as you go. Explore the depth and height of the new state and pump it. Rest into it and pump it. Allow yourself to float. Pump until ecstatic.

Repeat this a few times so that NOW becomes a distinct anchor and arises automatically from the gesture. Once you have done this, pump it up a few times and then test it as we did the other anchors.

Further Applications

NOW is, for most people an extraordinary experience. In its construction we bring together multiple feelings to create something new. What other combinations can you make? Can you discover a recipe for a feeling that you have never had before? What happens when you combine the anchors that you already have in different combinations? What might you add to NOW to increase its depth or value?

References for the exercise

Damasio, A. R. (1999). *The feeling of what happens: Body and emotion in the making of consciousness.* NY: Harcourt.

Damasio, A. R., Grabowski, T. J., Bechara, A., Damasio, H., Ponto, L. L. B., Parcizi, J., & Hichwa, R. J. (2000). Subcortical and cortical brain activity during the feeling of self-generated emotions. *Nature Neuroscience, 3(10),* (October, 2000).

d'Aquili, E. G. & Newberg, A. (1998). Why God won't go away the neuropsychological basis of religions, or why god won't go away. *Zygon, 33(2),* (June 1998).

d'Aquili, E. G. & Newberg, A. (2000). The Neuropsychology of aesthetic, spiritual, and mystical states. *Zygon, 35(1),* (March 2000)

Gray, R. M. (2001). Addictions and the Self: A Self-enhancement model for drug treatment in the criminal justice system. *The Journal of Social Work Practice in the Addictions, 2(1).*

Gray, R. M. (2002). The Brooklyn Program: Innovative approaches to substance abuse treatment. *Federal Probation Quarterly, 66(3),* December 2002.

Jung, C. G. (1979a). *The Archetypes of the collective unconscious. (CW9i).* Princeton: Princeton Univ. Press.

Jung, C. G. (1979b). *Aion: Researches into the phenomenology of the Self (CW9ii)*. Princeton: Princeton Univ. Press.

LeDoux, J. (2002). *The Synaptic brain.* NY: Viking Penguin.

LeDoux, J. (1998). *The emotional brain.* NY:Touchstone.

Maslow, A. (1968). *Towards a psychology of being.* NY: John Wiley.

Maslow, A. (1971). *The farther reaches of human nature.* Penguin / Esalen

Newberg, A., D'Aquili, E., & Rause, V. (2001). *Why god won't go away.* NY: Ballantine Books.

Exercise 6
Pacing The Future

Presuppositions Underlying the Exercise:

Pacing the Future presupposes the need for seeding generalization of wanted behaviors into multiple contexts. In every case where new behavioral skills are learned, their value is limited to the level to which they are contextually bound. In previous exercises we have learned to create powerful altered states of consciousness, but if they remain relevant only in the meditation room or therapy chamber, their real life utility is limited. Rossi (2000) suggests that every strong affect has an effective life span of 20 minutes to two hours. So, we can expect effects to linger into other contexts. But what happens after the state runs down or we are confronted with real life stressors without free or practiced access to resources?

There is a growing literature on the utility of meditation, acupuncture and alternative practices in substance abuse treatment. The research points to positive effects but often the effects end with the termination of treatment (Margolin et al., 2002; Morel, 1996). It is the author's belief that this is due to the simple fact that the positive results obtained from these practices tend to be contextually bound. The effects of Yoga are often bound to time, place, postures and patterns of meditation. The effects of acupuncture may be stimulus bound to the practitioner and the office where s/he applies them. Other practices may depend upon the continuing support of a community that reinforces the values imposed. The crucial value of this exercise is the spread of the experience of choice and control into the real world as an endogenous experience with a strong internal locus of control.

Bandura (1996) suggests several important methods for seeding the generalization of positive behaviors to other contexts. One depends upon the strength of the response itself: powerful experiences generalize to other contexts. A second depends upon imagined association. A third involves practice of the behavior in the appropriate context. Throughout the exercises we have sought to increase their behavioral salience through the intensity of their reward value. Here we use the other two methods to seed generalization.

The exercise depends upon the existence, in the anchors, of strong affective states that possess a high degree of behavioral salience. It further presumes that because these behaviors increase personal efficacy in any context, their expression tends to be self-reinforcing. By practicing them in multiple contexts we hope to establish those contexts as conditioned and discriminative stimuli for the states themselves and for the behaviors of eliciting the states. The exercise explicitly presupposes that one can attach a positive resource state to a future state of need by using an imagined representation of that future need state. The principle is soundly rooted in behavioral techniques. In the literature of NLP the technique is called Future-Pacing (Andreas & Andreas, 1989; Andreas, & Andreas, 1987; Bandler, 1985, 1993; Bodenhammer & Hall, 1998, Dilts, 1993; Dilts, Delozier, Bandler & Grinder, 1980; Dilts, Delozier & Delozier, 2000; Gray, 2001).

Generalization is presupposed to be a crucial part of the mechanism of addiction and substance abuse. If a behavior works in one context, or provides reinforcement in one context, it will tend to spread to others. Strongly positive responses tend to spread further than the less powerful. It is an explicit presupposition of the Program that the capacity for the individual to exercise choice over feelings will generalize into other contexts and that this generalization can be seeded using imagined encounters (Schaeffer & Martin, 1969; Wolpe 1958, 1982).

Appropriate performance of the exercise should result in the experience of spontaneous episodes of applying the states.

The exercise easily lends itself to application to systematic desensitization.

Expected Outcome

Participants will practice the resource states on a daily basis. From the resource states they will identify and describe a series of day-to-day contexts where the resources will be useful. They will plan a series of appointments for accessing the resources and monitoring their progress through the day. Participants will obtain practice using the resources in multiple contexts. They will have experiences of increased efficacy with regard to mood and behavior. They will experience the utility of the states on a practical level in contexts beyond the training room.

Instructional Notes

The current incarnation of the exercise is derived from Milton Erickson's classic paper on Pseudo-Orientations in Time (1954). It requires the participants to begin by entering the resource state that they want to use. From that state, after pumping it up, they are asked to imagine several points during the day when the resource will be particularly valuable. This links the resource states to internal representations of the real-world contexts and the need states associated with those contexts; it makes their appearance more probable as a simple matter of classical conditioning. The exercise continues with a series of planned appointments throughout the day when the participant will stop, fire off the anchor, enjoy the state and take some time to appreciate how well they have done through the day.

The technique makes use of pattern interruption (Grinder & Bandler, 1975; Watzlawick, Weakland & Fisch, 1988; Cade & O'Hanlon, 1993) to interrupt the progress of the day with experiences of feeling good. This allows the participant to reorient their attention towards more positive ends. It further requires them to reset the affective tone of the day on a scheduled basis. Each appointment starts with a strong positive experience which sets up felt effects that can last two hours or more (Rossi, 2000).

It is crucial to emphasize that the scheduled evaluations are to be positive. For each appointment, enter the state and note how well you did. In the evening it is important to plan to feel good for the morning and to finish the day in a positive state. Participants are asked to make written notes for two reasons: 1) They are committing to the positive tone in a record that enhances its reflexive credibility. 2) It is a record that can be checked and so, leads to increased levels of compliance. 3) The act of writing requires a certain level of reflexive analysis that marks out the positive experiences of the day.

There is often a certain level of resistance to taking the time required. Remind the participants that these are very short breaks and can be done anywhere. They should also recall that this exercise represents a significant means of caring for one's self. It is as valid for them as a coffee or cigarette break is for those who employ them. Reemphasize the value of caring for yourself. What have you done for yourself lately?

In the assignment, challenge the participants who have used the anchors solely as meditative tools, to use them in normal consciousness—with open eyes. Remind them that the states will take on a character and intensity that is appropriate to the context. Suggest to them that they will enjoy the sense of mood changing as background to their activities. Discuss contexts where they might find the states especially useful.

Behavioral Standards

Participants will produce a daily record of situations where they have applied the anchors. They will be able to articulate changes in their subjective experience of those situations after using the techniques.

Meditation

Any of the meditations provided can be used with this exercise. At this point every session should begin and end with an access of the NOW resources. Take the time to enhance the state so that it provides a dramatic experience of the deep self and an attractive altered state. Make sure that the participants have lots of time to enjoy the NOW state. As they do, invite them to explore new dimensions of feeling and being.

Exercise 6
Pacing the Future

An essential element of personal growth is flexibility. Biologists speak in terms of the Law of Requisite Variety—the organism that has the most options in any given environment is most likely to survive. Part of our efforts in these exercises is to increase personal flexibility and to increase the number of responses available to us.

One of the pleasant discoveries that typically arise from the Program is the surprising variety of responses that are available through the use of the anchors. We began with a weak memory that provided us with access to a feeling. We then spent time using the brain's own strategies to enhance the feeling and the memory. When the feeling reached a certain level of intensity, we abstracted the feeling from the memory and created a powerful state of positive potential that expanded to fill a range of possibilities beyond the original experience. When we had developed a series of five such states we created a deep sense of Self that embodied a full range of possibilities.

In each step we increased the depth and range of choice and experience. In the last several steps we have emphasized that each of the feeling styles that we have experienced represents a range of possibilities. FUN expands to include, deep joy, exhilaration, wonder, curiosity and play. YES awakens learning, synthesis, wonder, discovery, surprise and release. All of the states are now capable of multiple expressions.

Up until now, we have mostly used the states as meditative devices to create powerful altered states of consciousness. Most of us have already begun to use the states in real life settings. Almost everyone can attest to the awakening of new choices and new felt options in the world. With this exercise we are going to intentionally seed our daily experience with the new possibilities provided by the anchors.

In the exercise that follows we will be choosing one specific anchor each day for practice and observation. This does not mean that you can't use the others, it just means that for the exercise itself, use one of the anchors consistently. Over the course of the day, please take the time to notice a few things:

- Notice that when you use them with open eyes, in real life situations, the anchors tend to come up subtly in the background as mood changers.
- Notice that the anchor tends to produce a felt sense that is appropriate to the needs of the moment: alert where alertness is needed, distant where distance is needed, prudent where prudence is needed, etc.
- Notice how all of them are characterized by a sense of centeredness and safety.

The exercise will take a few minutes every day at scheduled times of the day. You will begin and end the day with a state of your choice and visit it several times during the day. We would like you to consider these as important appointments with yourself. They are times we would like you to set aside, just for yourself. They don't take long but they can change the quality of your daily experience. Please keep a record for yourself on copies of the Positive Resource Day Planner or something similar.

Please use all of the anchors. Don't just rely on your favorites. If you have favorites, alternate them with your less preferred states. Discover how many choices you have available. Remember, the appointments are times for you to enjoy. Each of the appointments that you take is really a time for you.

Positive Resource Day Planner

Upon arising, ask yourself: How do I want to feel today? State it positively (as what you want) and remember that it must be under your personal control. Make it something that you feel from within, not dependent upon your external performance, other people or events. Use one of the anchored states to access that feeling.

Fire off the anchor and take time to enjoy it.

While still in the state, open your eyes and think of three things that you must do today where the anchored feeling would provide enough extra strength, positive attitude, or humor to change the experience. List them here:

1._____
2._____
3._____

Now, schedule a minimum of three breaks during the day when you will stop, access the state and appreciate your progress towards it during the day. Write them in below. For each day, choose different times.

During each break make a note on how good it feels and how the well you've done so far.

Time Comment on the day's progress

_____
_____
_____

In the evening, before going to bed, access the state. Notice how good it feels and note how well the day went overall. From this state, plan the state that you might like to access tomorrow. End the night drifting off to sleep as you enjoy the state.

Write down your observations here and on the flip side if you need the space:

Further Applications

The inability to spread what we have experienced in the therapy room or meditation hall into everyday life has been a major pitfall of many otherwise valuable practices. The ultimate utility of each state is related to how strong the experience is when you anchor and re-anchor it. Remember that every time that you pump the anchor up, you are essentially re-anchoring it on a higher, more intense level. Whenever you use an anchor in a situation, you are linking the feeling associated with the anchor to that context so that you will be more likely to feel that way or use that anchor the next time you are there.

Imagine how your days would work if you took a few minutes to plan an anchor schedule to help you call up the resources that you need in the situations that you anticipate during the day? Each day, look through your schedule and find the places where different anchors would be most appropriate. Just before you enter each situation, take a few minutes to fire off the anchor.

References for the exercise

Bandler, R., & Grinder, J. (1979). *Frogs into princes*. Moab, Ut: Real People Press.

Dilts, R., Delozier, J., Bandler, R., & Grinder, J. (1980). *NLP. Vol.1*. Capitola, CA: Meta Publications.

Dilts, R., Delozier, J. A., & Delozier, J. (2000). *Encyclopedia of systemic neuro-linguistic programming and nlp new coding*. Scotts Valley, CA: NLP University Press.

Gray, R. M. (2001). Addictions and the Self: A self-enhancement model for drug treatment in the criminal justice system. *The Journal of Social Work Practice in the Addictions, 2(1)*.

Gray, R. M. (2002). The Brooklyn Program: Innovative approaches to substance abuse treatment. *Federal Probation Quarterly, 66(3)*, December 2002.

Exercise 7

Taking Control

Presuppositions Underlying the Exercise:

Taking Control presupposes that the participants have had significant experience with anchoring and applying the emotional choice that it provides. That experience will allow them to identify positive resources that they would like to experience more often, or places in their daily lives where they could use a state of their own choosing. It presupposes that they can access an appropriate resource, anchor it to a gesture or word of their own choosing, enhance it and have it available in a circumstance of their own choosing.

This presupposes once more that the more choice, the more options, and the more control a person has over their intra-personal environment, the less likely they are to return to problematic behaviors. It further assumes that by exercising control through the entire process, there will be further gains in perceived self-efficacy and self-esteem.

This exercise is assumed to be another milestone in the Program. It is the first fully autonomous activity using the skills from the Program. It is assumed to be a place where participants awaken on a significant level to the fact that these are their skills and theirs alone.

Expected Outcome

Participants will be able to identify a positive resource of their own choosing or a problem state for which none of the states already experienced meets the need. They will identify the state or feeling necessary, identify a past resource state that supplies it and create an anchor stimulus. Having created that anchor, they will then apply it to the problem state or enjoy it in its own right.

Instructional Notes

This exercise forms a review of the entire Program to this point but it places control firmly into the hands of the participant. From the outset, all of the choices belong to the participants.

In general, we prefer that participants choose positive states without regard to specific problems but need states often generate significant stimuli for creativity. Remind them to be imaginative and to use the examples others may provide. In general, creating positive resources is the better approach.

When participants have trouble finding states that meet their needs, work through them as class exercises. Poll the participants for their ideas. Work through a few in class and assign the remainder for homework.

For participants who have difficulty using the states with open eyes, invite them to engage with you in a conversation and fire off one of the anchors. Let them notice how the mood produced by the state arises gently in the background. Assure them that the state will conform to the context in which it is evoked.

Behavioral Standards

Participants will be able to choose a resource state from their own experience or create one to meet a specific need. They will be able to anchor it to a stimulus of their own choosing and future pace it to appropriate contexts.

Meditation

Any of the meditations provided can be used with this exercise. At this point every session should begin and end with an access of the NOW resources. Take the time to enhance the state so that it provides a dramatic experience of the deep self and an attractive altered state.

Exercise 7
Taking Control

Self-Efficacy is the feeling of being in charge; the feeling of being in control. From the first step, we have worked to build self-efficacy in various forms. We began with skills used to awaken and enhance memories. We continued with the creation and use of anchors. The feeling of self-efficacy was multiplied in the creation of NOW and spread into multiple life settings in Pacing the Future (Bandura, 1996; Gray, 2001, 2002).

When efficacy is experienced in multiple contexts, it gives rise to self-esteem. When self-esteem is connected to a deep and genuine experience of the Self, it becomes a powerful force for growth and transformation (Bandura, 1996; Gray, 1997a, 2001, 2002).

In any experience of personal growth one of the crucial passages is the discovery that the skills learned are truly your own: that they have nothing to do with a facilitator, a guru or a teacher; they flow from the center of your own experience and your own being.

Here is a crucial passage: throughout the preceding sessions, you have followed instructions on how to get certain effects. You have learned to access and enhance memories. You have learned to create anchors. You have learned to stack anchors creating brand new experiences and you have learned to use the anchors in real-life contexts. Now you are going to choose some states for yourself and create anchors that you find useful.

This is important because it places all of the control and all of the responsibility in your hands. You will choose the state and identify the resource. You will enhance the state and anchor it. You will test it, refine it and use it. It will be yours and in creating it you will realize that all of the skills that we have practiced are also yours and have been all along.

The Exercise

Create five new anchors for five states of your own choosing and then practice using them in the appropriate places in your present experience.

Here are some ways to find inspirations:

1) Think back over your life and find a memory that is characterized by a positive emotional state that you could use every day.

2) Find a state that you have read about or heard about. Find a way to experience it, taste it or imagine it.

3) Find a specific problem state that none of our anchors (FOCUS, SOLID, GOOD, FUN, YES, NOW) quite matches. After identifying the behavior or context, ask yourself questions like these: "How do I need to feel in that situation?" "When have I handled similar situations well? What was my state of mind?" "Who handles these things well and how would they feel?" Find an example state, anchor it and try it out.

For each example choose a resource and enhance it as we did in Exercises One and Two. Create the anchors using the techniques from Exercise Three. Enhance them using the techniques from Exercise Four. Do this for five separate states.

If you have difficulty thinking of examples, enter the now state and from there imagine some new resource that would be fun or interesting.

One suggestion of a positive state is a time when you were in what psychologists call a 'flow state'. Flow is a time when you were working on a challenging project, you knew what you were attempting to accomplish, but you didn't quite know how. As you worked at it, time disappeared and you became totally focused on the task at hand. That kind of absorption can be very useful. This is a good one.

A problem state that people often use is insomnia. For this, think of a time when you were exhausted (a pleasant childhood memory might be best), a time when you hit the bed, felt the pillow and the next thing you knew it was morning.

Another positive state might be an experience of deep peace or serenity.

Further Applications

By now, anchoring should make perfect sense to you. Take time to create some outlandish anchors. Create anchors for second sight, for ecstatic experiences, for extraordinary pleasures and your felt sense during strange experiences. Anchor spiritual experiences and experiences of esthetic wonder. Anchor fullness and desire. Anchor deep peace and unstoppable energy. Anchor the feeling of seeing in dreams or of hearing unspoken voices. Anchor the feeling of dropping off immediately into deep sleep.

References for the exercise

Bandura, A. (1997). *Self-Efficacy: The Exercise of Control*. NY: Freeman.

Gray, R. M. (1997a). *Ericksonian approaches to the ego-self axis: Establishing futurity and a sense of self in addictive clients*. Seminar: Innovative Approaches to the Treatment of Substance Abuse for the Twenty First Century. St. Francis College, Brooklyn, NY. Published on the WWW at http://www.temperance.com/nlp-addict/articles.html

Gray, R. M. (2001). Addictions and the Self: A self-enhancement model for drug treatment in the criminal justice system. *The Journal of Social Work Practice in the Addictions, 2(1)*.

Gray, R. M. (2002). The Brooklyn Program: Innovative approaches to substance abuse treatment. *Federal Probation Quarterly, 66(3)*, December, 2002.

Developing Self and a Sense of the Future: Exercises 8-12

Introduction to Exercises 8-12: Positive Resources, Setting Goals, ImageStreaming, Sponsoring a Potential

In the first seven exercises we concentrated on providing basic conceptual and emotional tools. We provided direct answers to specific behavioral needs. We established a behavioral bias towards positive memories. We established a sense of self-efficacy by teaching choice on an emotional level. We overcame most of the resistance by providing strong positive affect and establishing a positive set of expectations towards the Program and the other participants. We also began to awaken a sense of a deeper Self, a continuing positive identity.

In this second series of exercises we will build on that sense of connection to a deeper Self. We will find a sense of direction rooted in that Self and use that direction to build anticipation of a positive future. In that future, we will discover outcomes that express both the potentials and deep competencies of the individual. The series is characterized by accessing the most positive aspects of the past and encounters with an anticipated future that can become an irresistible motivator for further change.

Here, the logic underlying the Program shifts firmly into humanistic and depth psychological realms. The first exercises worked to assemble a basic tool-set and to establish the preconditions for deeper change. This second set of exercises seeks to awaken a path of growth that is implied by the felt tone of most basic of experiences: childhood aspirations, meaningful occupation, experiences of ease, competence and self-esteem. These are woven together to create a sense of "Who I am" and "Who I always have been". In so doing we instantiate the words of Jung and Maslow "What a man can be he must be" (Jung, 1966; Maslow, 1942).

The logic of the exercise flows directly from Jung's idea of the Self and its realization in the alchemical metaphor: Every direction in a person's life contains the seed of what that person *can* become and *must* become. Like the alchemists of old, if we assemble our elements (personal experiences) in the proper order, they reveal the *Quinta Essentia*, the true meaning that lies at the heart of all creation and towards which all of creation yearns. In the life of the human it is the full, conscious realization of the innate potential of the psyche. For Jung, the process was called 'individuation', the process whereby the person develops into an individual. For Maslow, it was the process of self-actualization. In fully maslowian spirit, the entire Program strives to induce a series of peak experiences which will awaken or reawaken the path to a fuller, more fulfilling life; a life that will ultimately provide a positive counter to destructive behavior (Gray, 1994, 1997b; Jung, 1968, 1977; Maslow, 1970).

This series of exercises begins with Exercise Eight, Positive Resources. This exercise comes in the form of a simple inventory of experiences that will provide an affective summary of the general tone of the participants' most empowering life direction. For each of the levels, the Facilitator's job is to emphasize the most powerful positive aspect of the memory or aspiration, and to further ensure that positive events are accessed. These may be analyzed for unifying themes so that a conscious sense of personal identity and direction is evolved. More importantly, however, the individual elements are examined for affective tone which may be used in the following exercise to create a new set of anchors. At this point, each participant is encouraged to explore aloud the positive connections that unite each of the elements in each of the themes.

Exercise Nine (Positive Resources Revisited) originally called for the participants to make use of the anchoring skills so carefully developed in the first sessions. When used, the participants create a series of five stacked anchors, each one representing an affective tone developed in the questionnaire from the previous exercise: anticipation (what did you want to be?), pride, confidence, identity (jobs and roles that have been important to you), innate competencies (the things you do well or have always done well), inherent intelligences and motivated learnings (things you learned easily), and experiences of self-esteem (times you felt good about yourself). Once anchored, these are again collapsed with NOW into a more global state of personal identity and positive self-regard.

In recent versions of the program, this exercise has been de-emphasized. As the resources are awakened in the context of the NOW anchor and as the discussions in session proceed in the afterglow of an initial experience of NOW in the same session, we now assume that they are incorporated by default.

In Exercise Ten, Setting Goals That Work, participants are challenged to envision a series of future outcomes that resonate with the feelings from the NOW anchor and to find ways that would make that kind of feeling the characteristic feeling of their life. This exercise makes use of a common NLP motivational technique known as the Smart Outcome Generator' or the well-formedness conditions for outcomes. The purpose of the exercise is to develop and test powerfully self-motivating outcomes (Bodenhammer & Hall, 1998; Dilts, 1993; Dilts, Delozier, Bandler & Grinder, 1980; Dilts, Delozier, & Delozier, 2000; Gray, 2001; Linden & Perutz, 1998; Robbins, 1986).

Exercise Eleven, Imagestreaming into the Future, is taken from the work of Win Wenger, Ph.D. It invokes creative processes related to dreaming and takes advantage of the power of spoken intention to solidify the futures developed in the preceding exercises (Gray, 2001; Wenger, 1979).

Exercise Twelve, Sponsoring a Potential, provides a final encounter with one's future Self. It ends the Program on an initiatic note and provides an almost spiritual experience for the last sessions.

Exercise 8
Positive Experiences

Presuppositions Underlying the Exercise:

The present exercise is rooted in the Jungian idea that every resource, every psychological experience and every pathology manifests, in some manner, the organism's inherent pull towards individuation. James Hillman (1996) speaks of this developmental 'instinct' in terms of 'calling', the idea that every individual has a particular niche or place in the world to which he or she is uniquely called and for whom that niche represents the perfect place for the working out of that potential (Gray, 1979b, 2001, 2002, 2003; Hillman, 1977, 1996; Jung, 1979; Progoff, 1956).

Implicit in this Jungian view is the idea that, by bringing together the appropriate elements, in the right order, an encounter with the Self as a conscious experience of psychic wholeness is constellated. Once constellated, that experience will tend to become the guiding impulse in the life of the individual (Gray, 1994, Jung, 1967, 1968, 1979).

Maslow predicted that an essential part of moving towards wholeness and Self-Actualization would be the development of peak experiences. According to Maslow (1970, 1971), peak experiences lie at the root of all religion and discovery. Each such experience is encountered as uniquely valuable. Each gives meaning to life itself and proves it worthwhile. An important part of this exercise is to seed the individual psyche with peak experiences that are accessible and that can help to provide personal direction.

While similar aims have been essential background assumptions of the previous exercises, the provision of integrative experiences of personal value and direction are the explicit themes of this and the following exercises.

From the perspective of neuroscience this exercise assumes the existence of preexisting schemas or behavior patterns that are linked to a more primal, more authentic pre-addictive identity. By evoking these patterns and linking them to present experience we hope to reawaken a personal identity that has been occluded by substance use patterns.

This pattern of memory linkage has been described in hippocampal mechanisms by Richard Morris (2006). In an experiment with rats learning place associations for food, he found that well established schemas could incorporate newly learned elements into their own structure and so incorporate the new information into long term memory. In this case it is believed that the awakened positive memories will be incorporated into the, by now, long term memory schemas involved in creating anchored resource states. Similarly, it is believed that the pre-addictive positive schemas (positive self-regard, self-esteem, personal agency) will be enhanced through the incorporation of positive affects related to the NOW resource.

Expected Outcome

Participants will be able to identify three to five examples for each of the classes of positive memories requested. As in the other exercises, these will be positive, sober, legal, harmless examples. For each example they will identify the positive feelings associated with the experience at that time. They will also note common themes relating each experience to the others.

Instructional Notes

As with every session since its first creation, begin this session with an extended experience of the NOW anchor. The assignment itself should be passed out at the end of the preceding session so that most of the work has been done before the session begins.

This exercise requires a bit more interpersonal and intuitive skill on the part of the leader than many of the others. Once the participants have produced their lists, the Facilitator may wish to bring out the common positive themes that link the answers given by any one participant. Often, the themes are not related so much to the surface category of the area-- sports, mechanics, and artistic endeavors-- as they are to underlying elements of complexity, team work, individuality or physical prowess. In general, we are looking for broad themes, what Gregory Bateson (1979) called abduction, "the pattern that connects."

It is important to note that this is the first of a series of exercises that require conscious verbal participation. Although participation was encouraged in the first eight sessions, most of the time was spent learning the skills, practicing the skills, or experiencing the states. In this exercise we want the participants to express themselves. Because we have waited for so long to open up to relatively free discussion, the following foundations are already in place to support positive communications: 1. Positive set. All of the exercises have emphasized positive affect and the entire context has become associated with the experience of resources. 2. Strong positive group identification. All of the exercises have resulted in strong positive experiences that have created a group dynamic rooted in shared skills, shared experiences, and a level of trust and openness based on respect rather than exposure. 3. Enhanced self-esteem. All of the participants have by now developed a significant appreciation of their own worth and the worth of their peers. It has been our experience that the communications offered here are almost uniformly positive, encouraging and self-affirming.

The entire exercise may be worked through in dialogue with the individual participants. Often, as the participants begin to share, the session becomes filled with enthusiastic recitations of long forgotten experiences. There is often a chain reaction in which a memory recited by one participant sparks off a similar experience from others. In general the individuals are polled for responses until the chain reaction begins. It will sometimes be necessary to interrupt and to force the group on to the next category.

We ask for examples from three stages of life; childhood, adolescence, and adulthood in order to find developmental themes and commonalties. In each stage of life the same central Core Identity persists. By seeking examples from each stage, we can make the unities more apparent. Moreover, these examples may be used to enrich the NOW anchor by adding a significant temporal dimension and a direction for personal growth.

Childhood aspirations can provide clues to ongoing competencies and identity issues. In some it will be necessary to get beyond the early power issues and the uniform goals shared by most children. For others, these continue as life themes. "What I Wanted to be When I Grew Up" is designed to extract the kind of anticipation and dream fulfillment experienced by the original dreaming child. It bears no relationship to whether those dreams were actually fulfilled or not. It relates rather to the affective tone of the dream and the connections they provide.

In these examples, self-revelations are often useful. I describe wanting to be a Paleontologist and imagining not only that I would find dinosaur bones, but that just around the next bend in the fossil creek near my childhood home I would encounter the living creature itself. Fantasy is a legitimate part of these exercises. Find out what made that experience or dream important to the participant and focus mostly on the excitement that it engendered.

Roles and jobs are important because they bring together personal capacities with social activities. Whether paid or not, whether supported or not, have the participants recall times when they really felt special or especially satisfied in fulfilling that role. "Jobs and Roles Since Childhood" seeks to define the specific affect that made those situations important. Here as elsewhere we are looking for very specific examples with clear sensory-based descriptions of the feelings involved. All of the events elicited in this exercise should be narrowed down to specific times and places.

Things That You Do Well asks: What were you good at, and what did that feel like? What does it feel like to know that you're good at something? Things You Learned Easily asks the participant to become aware of more basic competencies for learning and skill acquisition. These speak to root areas of competency. What you naturally do well and learn easily speaks to the places where your skills are centered. We also specifically ask about motivated learning. Some things were learned easily because an appropriate motivation appeared (as when a teenager learns to play the guitar).

Times When You Really Felt Good about Yourself are intended to reflect all of the above possibilities and more. It goes for a deeper level of self-acceptance than the others. It might legitimately include religious or spiritual experiences, mystical experiences (experiences of union) of a less identifiable kind. In a real sense, we are turning to Maslow's idea of peak experiences as an example of how good you can feel. By making these available we also provide a context for integrating the other states. That is, if we have no other clue as to how the different examples fall together, these peak experiences can provide an affective pattern that can help to organize the others. They also can provide a guiding sense, or affective tone, that will aid the participants as they discuss the exemplars.

I often give the following examples: When I was much younger, probably seven or eight, I lived on a river. From the back of my yard I could see the sweep of the river for about a mile between my house and a place across town called Marine Park. One day, while looking at the scene, I realized that I could walk, by myself, along the riverbank and get to Marine Park without needing to cross any streets. I had a sense of wonder and empowerment.

Another time, when I was about four, I found myself wandering through the Sunday School building at our church. In the middle of the building was a darkened area with a curtained stage. To the right of the stage was a window through which a sunbeam shone. Seeing the sunbeam shining through the window, I decided that that was where God must be and with great anticipation, went over to greet God in the sunbeam. Unfortunately I quickly discovered that it was ONLY a sunbeam. I have nevertheless anchored that sense of anticipation as a very positive experience.

Behavioral Standards

At the end of the exercise participants will have identified a series of positive experiences (a minimum of three for each category). They will be able to articulate some of the general themes that transcend the categories. They will be able to express a positive sense of personal recognition regarding the themes that appear in their own lives. Each participant will be able to express in sensory specific terms the specific affect accompanying each memory.

Meditation

Future Perfect can be used to good effect here.

Exercise 8
Positive Experiences

There is a certain sense in which all of what we have accomplished to this point has been preparatory. In fact, all of the preceding exercises were added to give more depth and impact to this one.

As previously noted, one of our central goals is the awakening of a deep sense of who you are and who you've always been. In the last several exercises NOW has awakened some of that sense and some of the depth of positive self-regard that you deserve as a person. Some of you have already begun to experience a deeper connection to a positive past. In this exercise we will focus on that connection as we ask what has always been true of you and will always be true of you?

We expect that this exercise will deepen your experience. It will add a temporal dimension, a sense of who you have always been. It will reconnect you to positive memories as resources. It will also provide a sense of your own personal direction; who you are becoming as you grow into your own calling or place in the universe (Gray, 2001, 2002; Hillman, 1996).

In order to do this we will ask you to discover six examples, more if you like, for each of the categories of experience noted below. In each case, as usual, we are most interested in the feeling that accompanies the memory, rather than the memory content itself. We are also interested in how you experienced it then, in context.

For example, the first kind of memory that we are looking at is the memory of things you wanted to be when you grew up. These don't have to be serious. It could be pure fantasy. In fact, fantasy is best. What did you dream of being? How did you fantasize and how did that feel? Go for the times when your imagination went wild and you got excited about the imagined experience. Get a sense of the feeling of excitement or pride that the fantasy generated. In one group, two grown men got into a friendly argument as to whether Aquaman or Submariner was the greater hero. This is the level we are looking for. When you have identified the memory, get a sense of the feeling of excitement.

None of the examples should give rise to the "poor-mes" or regrets over lost opportunities. All of the examples should be positive memories of your own experience in that time. Experience it as you did then. Use NOW as a vehicle for remembering.

To use NOW for this purpose, access NOW and spend some time enhancing and enjoying it. When you have come to a deep and pleasurable level, imagine that you can open a window in consciousness through which you will be able to view memories of the kind we are looking for. I usually imagine it as a parting of clouds. Use whatever imagery works for you. Continue to pump NOW until you have found several. As you find them, continue to pump NOW as you go for others. When you have found enough, pump up NOW and allow the window to close. Come back and enjoy the new memories.

Instructions

For each of the following categories, discover six examples, as noted. Take your time. Find as many as you can. This exercise is designed to help you find your inner direction. Choose positive, legal examples.

<u>What did I want to be when I was a child?</u> What did you dream about doing as a child? How did you see yourself in the future? (What made you feel good about it? How did it feel? What was your second favorite role/goal? Your third? How did they all feel the same? What was the unifying thread in all of them?) Take a minute and write down your examples. For each one, write down what was significant about it. As the list proceeds, note whether there was a unifying theme. Use some of the outrageous examples noted above. Find six.

<u>Jobs and roles since childhood.</u> What are six jobs or roles that you have filled that were really meaningful for you? (What made you feel good about it? How did it feel? What was your second favorite role/goal? Your third? How did they all feel the same? What was the unifying thread for all of them?). Write your examples and name the theme that runs through them.

<u>Things you do well.</u> Pick two from childhood, two from adolescence and two from adulthood (What made you feel good about it? How did it feel? How did they all feel the same? What was the unifying thread in all of them?). There may be some that you had to work hard to learn, others may have come easily. Write them down.

<u>Things you learned easily.</u> These consist of any subject, skill, talent or ability that came easily to you or that you really enjoyed learning. Pick two from childhood, two from adolescence and two from adulthood (What made you feel good about it? How did it feel? How did they all feel the same? What was the unifying thread in all of them?). There may be some behaviors that you have left behind. That's OK too. Write them down.

Times you really felt good about yourself; even if it got ruined later, even if it was just for a second. Pick two from childhood, two from adolescence and two from adulthood (What made you feel good about it? How did it feel? How did they all feel the same? What was the unifying thread In all of them?). Think about things that made you proud, things that made you feel confident. Think about mystical experiences and experience of being complete. Write them down.

Further Applications

There is no end to the richness of the resources that your mind provides. Each memory set that you awaken and enhance connects to a whole line of personal experiences that you can explore and reconnect to. What are some valuable experiences that would be useful NOW? How much history in this felt sense can you find? As you enhance them, what new possibilities do they suggest for you NOW?

References for the exercise

Gray, R. M. (1997a). Ericksonian approaches to the ego-self axis: Establishing futurity and a sense of self in addictive clients. Seminar: *Innovative Approaches to the Treatment of Substance Abuse for the Twenty First Century*. St. Francis College, Brooklyn, NY. Published on the WWW at http://www.temperance.com/nlp-addict/articles.html

Gray, R. M. (1997b). Addiction and the nature of meaning: Reframing in substance abuse treatment. Seminar: *Innovative Approaches to the Treatment of Substance Abuse for the Twenty First Century*. St. Francis

College, Brooklyn, NY. Published on the WWW at http://www.temperance.com/nlp-addict/articles.html.

Gray, R. M. (2001). Addictions and the Self: A self-enhancement model for drug treatment in the criminal justice system. *The Journal of Social Work Practice in the Addictions. 2(1).*

Gray, R. M. (2002). The Brooklyn Program: Innovative approaches to substance abuse treatment. *Federal Probation Quarterly, 66(3),* December 2002.

Hillman, J. (1996). *The soul's code: In search of character and calling.* NY: Random House.

Morris, R. G. M. (2006). Elements of a neurobiological theory of hippocampal function: The role of synaptic plasticity, synaptic tagging and schemas. *European Journal of Neuroscience 23*(11): 2829-2846.

Exercise 9
Positive Experiences Revisited

Presuppositions Underlying the Exercise:

In harmony with presuppositions underlying the other exercises, the current exercise presupposes that felt experience forms a better foundation for change than rules, mantras or other intellectual formulae. While we encourage people to make a conscious synthesis of the patterns of interconnection uncovered in the last exercise, it is our purpose here to create another experience of these patterns that can be evoked at will. Once again, we are creating conscious control over a set of normally unconscious experiences. In this case we are creating a much more complex state which we will expect to be available more often and in differing contexts. Because it is a powerful and pleasant exercise we will expect it to generalize across contexts more easily.

This exercise is in many senses the octave of Getting to Now. There, we combined five simple states into one complex state. Here we combine several exemplars from each category into complex states and then stack those into what might be thought a hyper complex state that includes the original NOW anchor.

Our presuppositions include the idea that we are creating a rough equivalence to a Maslowian peak experience; one that can and will allow the individual to redirect his or her energy into more natural, Self-Actualizing directions. If persons involved in the exercise have a history of religious or mystical experiences, we will expect this experience to resonate with them (Maslow, 1970).

The exercise itself makes use of the same skills developed in part one; anchoring, enhancing states, stacking anchors and applying them to other contexts.

Expected Outcome

Participants will have the experience of establishing a set of six new anchored states.

1) An anticipatory state that assimilates their experiences of growing, developing and dreaming about future roles. 2) A state of positive identity and self-efficacy related to jobs and roles. 3) A state of confident ease about personal abilities. 4) A state of confidence in learning and knowing that "I possess certain competencies". 5) A state of feeling good about oneself. 6) The combined state of positive self-regard and personal direction compiled from the rest as they are added into the pre-existing NOW anchor.

The final state is expected to provide a powerful sense of positive self-regard not limited to circumstances or contexts. As such it may be used to outframe other feelings and identities.

Instructional Notes

In current use, this exercise is passed over in favor of time spent discussing the results of the previous exercise and enjoying the NOW anchor. Nevertheless, if used, the current exercise follows directly from the basic anchoring exercises with the significant difference that each finger is being used as a complex anchor for multiple experiences.

As with all of the exercises since the creation of NOW, begin the session with an extended experience of NOW.

To begin with, ensure that each participant has identified at least three memories or experiences from each of the five groups (things you wanted to be, jobs and roles since childhood, competencies, aptitudes and, episodes of positive self-regard). Ensure that each of the examples focuses on a strong positive state. Like the other anchoring exercise, each exemplar should focus on a specific memory or event. It should also focus on the best available instance of that memory.

In this exercise, the conditioned responses will be associated with the five fingers (thumb through pinky) of one hand. The hand should be laid flat on a table, arm rest, or other flat surface (stomach or thigh would do just as well). Each class of exemplar (things you wanted to be, jobs and roles since childhood, competencies, aptitudes and, episodes of positive self-regard) will be anchored to one of the fingers of one hand, in order.

For this exercise, begin by accessing the anticipation associated with the participants' most favored vocational dream or fantasy. Remember to focus on the positive feeling from the perspective of the child.

We have found it useful to start with one example and enhance it. As the homework asked them to clearly identify the feeling, participants should be able to quickly identify the location of the feeling and how it spreads to the rest of their body. They can then begin to spin it through its own center (see Exercise One and Exercise Four) until the memory fades and they are floating in the pure feeling state. From this state, suggest that they add the other memories for the categories, one at a time, and spin them into the state. Do this for each category; one category to each finger.

When the deep states representing each of the categories has been appropriately constructed and enhanced, create anchors as follows.

Access the first category by its first exemplar. Enhance the feeling to peak and begin to anchor this by tapping the thumb on the flat surface. Shake free and repeat. After two or three iterations, do the same thing, on the same finger, with the next memory and the next, until the entire category has been associated to the tap of the thumb on the flat surface.

Continue to anchor each category to a separate finger in the same manner, until all of the categories and their exemplars have been associated with a specific finger. Test and enhance each anchor by tapping that finger just as the pumping gesture was used in Exercise Three. That is, tap the finger and wait for a feeling to arise; just as you notice the feeling coming up, tap the finger again. As the enhancement evoked by tapping arises in consciousness (just at its first hint of its arising), tap again. Continue in this manner until each category state is fully enhanced. Advise the participants that new exemplars that enter consciousness spontaneously may be anchored in as well.

When the participants have enhanced each of the category states associated to the appropriate fingers, let them stack the states from each of the fingers into the NOW state as follows. Tap the thumb. As the thumb-state arises into consciousness, take a few minutes to become aware of how it arises in consciousness. Notice the patterns of tensions, relaxations, temperatures and movements that characterize the state. Hold them in mind as a pattern and fire off NOW. As NOW arises into consciousness, become aware of the way it merges with and compliments the thumb state. Become aware of their interplay and how NOW integrates the energy. Become even more aware of their combined pattern as you tap the first finger. As that state arises into consciousness, notice how it relates to the new patterns in NOW. Notice how it arises and what patterns of tension and relaxation, movement and stillness it creates. Focus more on its specific characteristics and as you do, fire off NOW. Continue cycling through in this manner until all of the fingers and the states associated with them have been integrated into the NOW state.

As the new experience continues towards peak, begin to pump the NOW anchor while focusing on the new qualities of the experience. Let the participants continue to cycle through the exercise until NOW fully integrates the newly anchored material.

Behavioral Standards

At the end of the exercise participants will be able to express their perceptions of the new state. It will, as an emergent property of the interactions of the other states, be unpredictable from the simple sum of their parts. In general, it is a strongly positive feeling of self-esteem that carries with it a sense of personal direction and continuity through time. Participants should enjoy the state and be pleasantly surprised by its familiarity, its intensity and its newness. Participants who have not done the preceding exercises will not be able to create the new state.

Meditation

Future Perfect is used following this exercise. This sets the groundwork for the following exercise and reintroduces the participants to a future-self grounded in positive experiences from their entire life span.

Exercise 9
Positive Experiences Revisited

By now, you have assembled quite a list of positive experiences. There are probably many that surprised you as they came to mind. Part of the purpose of the exercise was to connect you with the richness of your own experience. One of the basic presuppositions from NLP is that every person has a rich trove of positive experiences that they can use as resources in the present.

There is, however, a more fundamental idea that comes from the works of Carl Jung. Every experience, whether positive or negative, carries within it a tendency to point towards your deepest center; the core of your Self. This tendency is part of all of the feelings that we have. On one level it is just obvious that we are the same person having the feelings. On another level, less obviously, all of our experiences encode our deepest needs, desires and aspirations. When properly assembled, they begin to awaken us to our life calling and the meaning that so often eludes us.

Use the following instructions to create a new set of anchors and to add them into your experience of NOW.

Instructions

1) Use the lists from the last exercise and for each memory in each category to access the anticipation or other good feeling from the memory. Begin with Things You Wanted To Be. Step into the first example from your list and become aware of the feeling it generates and where that feeling arises in your body. (If you cannot find the feeling, use the techniques from exercises one and four to enhance the memory to the point where you can really feel it.) Begin to rotate the feeling faster and faster until the memory fades and you find yourself floating in the pure feeling state. Enjoy the state for a few minutes and then begin to add the other examples from the same category (You may not even need the list). Do this by gently turning your attention to the other examples, or by opening a window in the state to access the memories. As each comes to mind, notice what it adds to the present state and spin it into the feeling. Do this for all five categories.

2) When you've gone through all of the states and all of the categories, create anchors for each of the states. Use one hand and the fingers of that hand in order, as anchors for each category. Use your thumb for things you wanted to be, your pointer finger for jobs and roles since childhood, your middle finger for things you did well, your ring finger for things you learned easily, and your pinky finger for times you felt good about yourself.

Start with your first example of things you wanted to be. Enhance the feeling and begin to anchor this by tapping your thumb on a flat surface. Shake free and repeat. Make sure that your tapping is consistent in rhythm and intensity. As you feel that the anchor is beginning to work, add in the other memories, one at a time. Do this by holding the felt sense of the memory in mind and body and gently shifting your attention to the next experience. Spin each till the memory disappears and tap some-more. When you've added all of the examples into the anchor and you feel that the anchor is working well, shake out the feeling. Then, move on to the next finger and the next category until every finger has become an anchor for one of the categories.

3) Test and enhance each anchor by tapping the finger. Use the tapping movement just as you used the pumping gesture from Exercise Three. That is, tap the finger and wait for a feeling to arise; just as you notice the feeling coming up, tap the finger again. As you notice that tapping increases or changes your experience (just at the first hint of its arising), tap again. Do this with each finger until you've finished with the lists and all five fingers. If new examples come to mind during the process, feel free to add them in.

4) When you have completed anchoring and testing the new states, add them into NOW as follows. Tap the thumb. As the thumb-state comes into consciousness, take a few minutes to become aware of how it feels. Notice the patterns of tensions, relaxations, temperatures and movements that characterize the state. Notice its speed and texture. Hold all of these in mind as a felt pattern and fire off NOW. As NOW arises into consciousness, become aware of the way it merges with and compliments the thumb state. Become aware of their interplay and how NOW integrates the energy. Pump NOW as you pay attention to their interaction. Hold these new feelings in mind as a pattern. Tap the first finger. As that state arises into consciousness, notice how it relates to the new patterns in NOW. Notice how it arises and what patterns of tension and relaxation, movement and stillness it creates. Focus more on its specific characteristics and as you do, fire off NOW. Pay attention to how NOW integrates the feelings...

Follow this pattern, finger, NOW, finger, NOW, finger... cycling through all of the fingers so that all of the new states are integrated into NOW. As the new experience continues towards peak, begin to pump NOW while focusing on the new qualities and new dimensions of feeling that are now part of NOW. Go back through the cycle several times and discover new things about you and about NOW.

Further Applications

Besides being used to enhance NOW, you have created another useful set of anchors. Implicit in this exercise is the flexibility of NOW. What stream of felt experience would add an especially useful dimension to NOW, now? What new directions of skill or perception would you like to be more available?

As you enjoy the new depths of NOW and the directions that they imply, can you find positive themes that stretch all the way back as far as you can imagine? What happens when you integrate them into the experience?

References for the exercise

Gray, R. M. (1997a). Ericksonian approaches to the ego-self axis: Establishing futurity and a sense of self in Addictive Clients Seminar: *Innovative Approaches to the Treatment of Substance Abuse for the Twenty First Century*. St. Francis College, Brooklyn, NY. Published on the WWW at http://www.temperance.com/nlp-addict/articles.html

Gray, R. M. (1997b). Addiction and the nature of meaning: Reframing in substance abuse treatment. Seminar: *Innovative Approaches to the Treatment of Substance Abuse for the Twenty First Century*. St. Francis College, Brooklyn, NY, Published on the WWW at http://www.temperance.com/nlp-addict/articles.html.

Gray, R. M. (2001). Addictions and the Self: A self-enhancement model for drug treatment in the criminal justice system. The *Journal of Social Work Practice in the Addictions, 2(1)*.

Gray, R. M. (2002). The Brooklyn Program: Innovative approaches to substance abuse treatment. *Federal Probation Quarterly, 66(3),* December 2002.

Hillman, J. (1996). *The soul's code: In search of character and calling.* NY: Random House.

Peck, M. S. (1998). *The road less traveled: A new psychology of love, traditional values and spiritual growth (Third Ed.).* NY: Simon & Schuster.

Exercise 10

Setting Goals That Work.

Presuppositions Underlying the Exercise:

Setting Goals That Work is a fairly standard NLP model of highly effective strategies for motivation and outcome planning. According to the originators of the model, good decisions follow syntax much as languages do (Much of NLP is rooted in chomskian linguistics). According to that model, this exercise represents an operationalization of the syntax of good decision making. In its present incarnation it can serve as a decision evaluator and as a motivation enhancer (Bodenhammer & Hall, 1998; Dilts, 1993; Dilts, Delozier, Bandler & Grinder, 1980; Dilts, Delozier, J. A. & Delozier, J., 2000; Gray, 2001; Linden & Perutz, 1998; Robbins, 1986).

For decisions that are appropriate to the individual and their long-term goals, the technique builds a multilayered, multi-sensory representation of the outcome that, for many, becomes quite irresistible. In those cases where the outcome is inappropriate or ill formed, the participant experiences the outcome as clearly inappropriate.

The fundamental presuppositions are rooted in Jung's position that response systems which are rooted in archetypal structures tend to become compelling. A correlate of that presupposition is that the Self, as the archetype of wholeness, can become the most potent of motivators. By rooting the exercise in the sense of Self developed in the NOW anchor, we create an outcome that resonates with deep levels of individual structure (Hillman, 1994; Jung, 1979; Progoff, 1959)

The exercise follows Bandura's (1997) observation that motivation is tied to efficacy beliefs. People are more likely to choose behaviors, expend effort and persist in the face of failure if outcomes are tied to a strong sense of self-efficacy. Consequently, the more grounded the outcome is in the client's experience of competence, the more powerfully motivating it will become.

The exercise is more directly derived from Milton Erickson's Pseudo-Orientation in Time (Erickson, 1954). There, he sets forth the observations that effective hypnotic interventions represent the skillful manipulation of past resources. He further differentiates between empty fantasies which have no root in personal experience and unconsciously directed goal-oriented behaviors that are founded on personal potential.

For the most part, these characteristics are typical of deep, intrinsic motivations. Intrinsic motivations are desired for their own sake. They are meaningful to the individual independent of external pressures or rewards. They are contrasted with extrinsic motivators which include things like money, sex, power, fame and popularity: stuff. Extrinsic motivators are well known for their capacity to sometimes weaken intrinsic motivations. When, however, they are simply the fruit of a deeply held personal direction or outcome, they present no such problem (Deci & Ryan, 2008; Hulleman et al., 2008).

Intrinsic motivators are desired positively (Deci and Ryan, 2008; Gray, 2005, 2008). They are characterized by choice and personal autonomy; they often include strong self-efficacy beliefs (Baumeister & Heatherton, 1996; Deci & Ryan 2008; Hulleman et al., 2008; Koestner, 2008; Nootz, 1975). Because they are often rooted in previous or vicarious experiences, they can be specified in sensory terms (often with special emphasis on kinesthetic elements—this is how I will feel) (Baumeister & Heatherton, 1996).

This exercise is rooted in several presuppositions from NLP that are non-intuitive.

1) There is a syntax to internal representations that differentiates between well-formed, self-motivating outcomes and ill-formed or inappropriate outcomes. 2) That same internal syntax differentiates between motivating outcomes and transformative/generative outcomes. 3) Negative outcomes ("I don't want to be X any more") are, by definition, ill-formed. 4) Motivation is keyed to control. An outcome that is not under the direct personal control of the participant will not produce appropriate levels of motivation. 5) An imagined experience of future success can generate real-world guidelines for attaining a goal. 6) Properly imagined outcomes are constructed from present time competencies and serve to enhance the value of those competencies (Andreas & Andreas, 1987, 1989; Bodenhamer & Hall, 1998; Erickson & Rossi, 1980; James & Woodsmall, 1988; Linden & Perutz, 1998; Miller & Berg, 1995; Robbins, 1986).

More intuitive presuppositions include: 1) The more completely an outcome is represented internally as sensory information, the more compelling it becomes. 2) The more that a representation of a future outcome is integrated into one's life in terms of its impact on current habits, abilities and relationships, the more real and compelling it can become (Bandler & Grinder, 1979; Bandura, 1997; Erickson & Rossi, 1980; Grinder & Bandler, 1975a; Robbins, 1986).

Continuing the theme of constellating the Self, this exercise requires the participant to visit five separate futures in the realms of spirituality, relationships, intellect, occupation / work and health. Using the

Ericksonian Pseudo-Orientation in Time to activate potentials in each of these areas, we also foster the awakening of a deeper sense of Self.

Prochaska hints at the importance of future orientation and its foundational relation to the dynamics of change. He observes (Prochaska, et al., 1994) that the key to movement from *precontemplation* to *contemplation* in the Stages of Change Model is the awareness of a future goal that is personally more valuable than the problem behavior. In his study of the relationship, the positive valence of the desired change increases by one full standard deviation while the perceived valuation of the problem behavior decreases by .5 standard deviations. The change in attitude towards the desired future predicts all of the change towards the unwanted habit. It also predicts most of the overall success in treatment (Prochaska, et al., 1994).

An essential component of the exercise is the differentiation between transformative outcomes and simply motivating outcomes. An outcome can be made motivating insofar as it meets the well-formedness constraints listed in the exercise. However, even though the outcome may be motivating, it will not empower generative transformation unless it is rooted in the deep Self and the personal directions that it defines. As a result, in this exercise it is crucial that the chosen outcomes emerge from the felt sense of the NOW anchor. This will ensure that they carry the individual into the path of transformative motivation that we have been discussing.

Expected Outcome

Participants will select an outcome based upon their experience of the NOW resource. They will gain personal experience in applying the syntax of goal setting that will allow them to use the technique at other times. They will understand that the strategy may lead them to abandon an inappropriate or ill-formed outcome or become powerfully motivated to pursue an appropriate, well-formed outcome. They will have an experience of encountering themselves in a positive future environment.

Instructional Notes

As you present the exercise you may want to share with the group how often we all have been accused of lacking motivation; how everyone talks about motivation, but no one tells you how to get it. Let them know that this exercise is a motivation builder. Motivation is not something you have; it is something that you do. Motivation is about having positive expectations that are real enough and reasonable enough to make you want them. Another important part of motivation is having it match your personal values. In this exercise we will learn how to evaluate and install motivating futures. Again, there is an implicit message about self-control and personal autonomy. Motivation is no longer a gift from heaven but something that each of us can construct for ourselves.

A certain part of this exercise should be used to explain that the brain is unable to distinguish between imagined and experienced realities. Much like autogenic training, we are building a reality that our brain can use to practice the reality which we want to attain. The more really we represent the goal state, the more fully we can experience it, the more motivating and empowering it will become (Cade & O'Hanlon, 1993; Damasio, 1999; Erickson & Rossi, 1954; LeDoux, 2002; Linden & Perutz, 1998; Robbins, 1986).

The power of visualization was illustrated by the Russian Olympic teams who, in the late 1980s, were able to show that visualizing successful performance transformed itself into a significant amount of equivalent training (Schroeder & Ostrander, 2000). Other research using fMRI to observe the functioning brain has shown that, with the exception of intensity measures, imagined activities and "real" activities are largely indistinguishable from measures of brain activity. That is, the same neurons fire whether the practice is real or imagined (Iacoboni, Woods, et al., 1999).

The exercise is structured so that the technique can be used for any goal or outcome that the participant can imagine. For the purposes of the Program at this stage we prefer that you emphasize outcomes that resonate with the NOW state. Explain to the participants that they will be expected to work only with outcomes that emerge from the NOW experience.

The exercise begins with the elicitation and enhancement of the Now experience from the previous exercises. This forms an intuitive direction for choosing an appropriate outcome. Suggest that the participants use that feeling to guide them to a future which would allow them to feel that way every day. What would they need to be doing? What would they need to change in their life? What would they need to add to their present experience to make this feeling the characteristic feeling of everyday life?

This part of the program takes advantage of the well-known NLP criteria for well-formed outcomes. In this case, however, instead of evaluating an outcome or making an otherwise lackluster outcome powerfully motivating, we are beginning with a felt state of positive personal identity generated by the NOW state. The question that drives the exercise is 'If you could feel more like this every day, what are the relationships, jobs, spiritual activities, health behaviors and intellectual pursuits that would support it? What things do you already do that tend to make you feel more like this?'

As importantly, it makes use of Prochaska's strong principle of change. James Prochaska, one of the originators of the Stages of Change Model discovered that when people are changing, the most important element in that change, the thing that will drive that change towards successful completion, is a future outcome that is more important than the problem behavior. NLPers have known this for a long time. But now scientists and mainline psychologists understand it too.

It begins by firing off the stacked resource state (NOW) and creating an inner experience (one at a time) for each of the following categories. Instruct the clients to do this by opening a window in their imagination and stepping out into the future.

As you lead the exercise, instruct the participant to go into the future to discover an outcome--not with a plan in mind. Let the plan or outcome emerge from the experience of NOW. For each category, instruct them to experience what they are are doing, not what they have; stuff is not an appropriate outcome.

For each kind of life experience invite the participants to do the following:

Fire off the state and pump it up. As you enhance the state, imagine that you have been using these kinds of states for the last five years and your life has changed because of it.

Allow yourself to travel into a future characterized by this kind of feeling, this kind of being. Open a window in your imagination and begin to notice what must be true about your life if this feeling has been the growing center of your life and the things that you do. Become more and more aware of what it is that you are doing, NOW… and how it feels in this future. As you journey into that future, take note of the positive evidences of the things that you are doing.

In separate instances or even during separate sessions, have the client apply this frame to the categories of experience listed below. Encourage them to stay with it until they get a concrete sense of the kinds of things that they are doing in that future for each category. Many clients will find the realities growing piece by piece with successive returns to the NOW state. Encourage them to go through the different domains, one at a time. The order is not important. Let each participant go through them in the order that is most comfortable to them.

1) Spiritual life. When you get in touch with who you really are, what kind of spiritual aspirations do you have? Who would you emulate? What would you do differently? How would you feel? How does this relate to or emerge from the resource state? Use NOW to experience your spiritual life in the future. Notice how you give physical expression to this reality. Notice the kinds of things that you are doing, especially the things that increase and deepen your ability to grow as an individual in a spiritual context.

2) Relationships. When you fire off your resource state and think about your future relationships; what will you be doing differently? What will have changed in the way you respond to others and they, in turn, respond to you. How does this relate to or emerge from the resource state? Notice how you give physical expression to this reality. Notice the kinds of things that you are doing, especially the things that increase and deepen your ability to grow as an individual in relationship to others.

3) Intellectual life. Fire off your resource state and think about how you use your brain and your mental capacities. How are you using them differently? What will you be doing more? What will you be doing less? What challenges you and how do you enjoy it? What is your next level of educational need? How do you feel? How does this relate to or emerge from the resource state? Notice how you give physical expression to this reality. Notice the kinds of things that you are doing, especially the things that increase and deepen your ability to grow as an individual and in your ability to use your intellectual abilities and gifts.

4) Occupation/work life. Fire off your resource state and ask yourself what you must do (or do more frequently) that will allow you to feel this way more often. Do you have a different job? Have you started a new career direction? Do you work less? Have you started a hobby? What is different? What are you doing in this future and how does it make you feel?

5) Health. Fire off your resource state and ask yourself what you must do (or do more frequently) that will allow you to feel this way more often. Have you changed jobs? Have you changed eating or exercise patterns? Do you work less? Take more vacations? What is different? What are you doing in this future and how does it make you feel?

After your client has visited each of these futures, have them write down one or more alternatives that immediately come to mind. Some clients will immediate settle on one outcome, others will have a few. Writing them down will help them to focus on the more motivating options.

Use the following procedures to test and concretize their outcomes. You can talk them through these or provide a written list. If any of the steps is unclear or difficult to resolve, have them fire off NOW and allow the answers to emerge from within the state.

1. <u>State your experience of that future as a positively desired outcome. Express it in terms of behaviors: what you will be doing.</u>

Think about what you have experienced in your vision of the future. Begin to describe it as a positive thing, something you can hold in your hand or put in a wheelbarrow or something you can do.

I will get my Associate's Degree is a good outcome. You can see yourself at graduation.

I will begin doing more of my art, my music, my… . I will spend more time using deep states to access new capacities. I will be more familiar with my internal landscape and will have new ways of generating options. I will have new choices in how to spend my time and how to focus my attention. I will finally be able to….

2. <u>Make sure that it is under your personal control.</u>

A proper goal must be under your control. It must be something that you can do: saving money to buy a house or business, getting the training and connections to make a career change, finding new ways to change the way you think or behave, leaving an old relationship or starting a new one. These are all things that you could do. All of these things are choices that would be under your control.

Under your control implies that it does not depend upon someone else or some external agency. It is something that flows from your own capacities. If it depends on someone else's action, it is probably a pipe dream.

As you focus on the NOW state and how you will be feeling after using it for five years and growing with it, what new sense of your own capacities arise? What do know that you can do, because you are feeling more like this every day?

Personal control also includes reasonableness. Is the goal you want realistic, or should it be broken up into sub goals or outcomes?

3. <u>Specify three different ways in which you will know that you've gotten it when you get it.</u>

How will you look when you have it? How will you feel? Who will be there? What will you see and feel? The more fully you can imagine getting it or doing it, the more powerfully motivated you will be to get it. Use all of your senses.

Fire off NOW and imagine that you are there and you can see it and feel it and touch it. It is really important in this step to feel and see and hear and taste and smell your future reality. The more senses you can use the more motivating the goal will become. The motivational centers of the brain encode preference and motivation in terms of full, multisensory representations.

If you're thinking about a restaurant business, envision the deed or licenses, hold them in your hand, feel yourself signing the deed, writing the menu, opening the front door on the first day of business. Smell the food cooking in your ovens.

Feel the handshake from the boss and see his face as you are given the promotion.

Smell the fresh ink on that new book. Feel its heft in your hand and the crispness of the pages. See your name on the cover and on the spine.

See the body language and hear the tones that tell you that that relationship really has changed. Notice how you hold yourself and the people that you love. Notice their responses to the new you.

Feel the exhilaration as you cross that finish line or reach that peak of physical activity. Notice the positive change in your muscle tone, your energy, your breathing. Notice the change in your gait and posture. Discover new kinds of personal satisfaction.

See the changes in your face, posture and movements as you live out new spiritual realities. Become aware of the depths and directions of feeling and experience as you awaken into that new reality.

For spiritual outcomes, fire off now and notice where that leads you spiritually. What kinds of practices behaviors or realities open to you and support this direction of personal growth? How do the future experiences impact the depth and richness of your current experience? As you focus on the practice of using the anchors, how does your experience after five years feed back into the way they work now?

For this step, specify three of these kinds of things that will really let you know when it's yours.

4. <u>Decide where and when it will be appropriate. Will you want this all the time? Is it appropriate everywhere? Should it be limited to a specific context?</u>

As we make a goal realistic, it is important to understand that it may not be appropriate at all times and places.

Fire off NOW and consider where this fits in your broader circumstances. How does this outcome fit in your life priorities from the perspective of NOW? What kind of priority does it have and when will it be good to work on it? From the perspective of NOW, are there other things that are more important or pressing? How does this integrate or resonate with the other changes you are planning?

5. <u>What will it change in your life and in the lives of the people around you?</u>

Real goals have real consequences. When you are sober, you will have new friends and new relationships. How will this work for you? Who will support you? Who will resist you? What will it cost in terms of relationships? What will it get you? Are you willing to take the chance? What will you gain in terms of new opportunities?

When you enter a new business or a new neighborhood, there are also changes in your life. How many hours will you be working now? How many nights, how many weekends? From the perspective of now, begin to experience with a sense of appreciation, the kinds of changes that you will be experiencing in the world around you and notice how well your new perspective prepares you for them.

6. <u>Experience now, in your imagination, how you will look and feel, what you will see and hear when this is a reality.</u>

Use NOW to go back and get the image again. See your surroundings in bright color. Experience the people around you. Experience in full sensory detail, how you will feel and how those around you will respond. Step all the way into your own body and experience the future from inside.

What we are accomplishing here is getting in touch with your future self; the self who has already accomplished your goals. It is important to feel and identify with this future you because he or she will show you how to get where you want to be.

Use NOW to fill out the details. With each pump, begin to notice how more and more detail arises into consciousness. Go through all of your senses. Step all the way into it. See it, feel it and hear it from your own perspective. How do you feel having it? How do you hold yourself? Move into that same posture. What do you say to yourself? What do the people around you say?

7. <u>Move backwards from the final realization of the goal to discover the steps that make it possible.</u>

Once you have a real experience of yourself having what you want, take time to get into it. Enjoy it. Feel it and continue to feel it. Then, from that place, where you can see it and feel it and taste it and hear it, look back towards today and find the steps that got you there. Begin with the last step, the finishing touch. Really be there. Ask yourself "What was the last step that I made that put it together?" Take your time. When you've answered that question, ask: "What was the step before that? … and before that?" Take the time you need to find the steps that got you there. Keep the state, feeling like you've already gotten it. This will guide your mind to the steps that got you there.

Make sure that there is a manageable number of steps and not too many. Ask your unconscious to help you to make the size of the steps just right.

For some of the Domains and for some of the clients, the exercise does not immediately and cleanly end in an outcome that can be worked through the well-formedness conditions. This is often true in the case of clients who have gone into so deep a state of trance that they emerge amnesic for most of the process. The clients often emerge with a strong felt conviction that things will change and be better, or they will couch their experience in nominalizations and unspecified language.

In such cases, the facilitator should probe the client with metamodel questions until concrete objectives can be clearly stated:

If this is how you feel in this state, in the future, what kinds of things will you be doing that will support this feeling and the way of life that proceeds from it? Think of some things from your past, in this same domain, that have created similar feelings. Name some of them and tell me whether they fit with your unconscious vision for the future. Do you know other people who you think felt this way? If you do, how did they respond in this area that was different from the way you used to do it? Step into that—their pattern—try it ot on and test whether it works for you.

For each domain where the client returns from NOW with no concrete objectives, press them in this manner until they can state a concrete set of objectives for that future life. Once you have done this, work them through the Smart Outcome Work Sheet.

Another useful technique in this situation is John Overdurf's (2006) end state energy pattern. Ask the client to reassociate to the NOW state in that specific domain (work, spirit, relationships, health, intellect) and from within that domain, as seen through NOW, ask them what is the smallest next step that they can take that will begin the process of manifesting those realities in their own life. The response may seem to be irrelevant, but as long as it is congruent with the felt state of NOW in that life domain it is acceptable.

8. List the five steps necessary to get from here to there.

Think about the steps that you just learned from the future you. Break the list into five steps that you can handle. If necessary, the five steps can be five sub goals and you can use the process on each of the sub goals.

After working through each of the domains, spend some time discussing them with the clients. In a group setting the discussion will help the other participants to work through any parts of the process that may have found challenging.

Behavioral Standards

At the end of the exercise, participants will be able to define well-formed goals for the contexts of Spirituality, Relationships, Intellect, Occupation/Vocation and Health. They will be able to describe five or six concrete steps towards achieving them. For the purposes of this exercise each goal should reflect some new behavior that will enrich or transform their experience of life.

Trance / Meditation

The Future Perfect meditation is used with this exercise.

Exercise 10
Setting Goals That Work

One of the important aims of the Program is to create a goal that resonates with your deep sense of who you are and what it is that you are becoming. In the last exercises we added a connection to your past to give some depth and direction to NOW. By this point you will be experiencing yourself, not only as centered, controlled and feeling good, but also as someone who has a future. Your life has meaning, your life has direction. There is a future that awaits you.

There are several ways to describe this path. Biologists might call it a developmental pathway or chreode. Spiritual and religious people might think of it in terms of a calling. Secularists might just like to think of it as the call of your biology to the ecological niche for which you are best suited. In general, we like to think of this exercise as a kick-start on your way to finding the place in the world that is best for you. Joseph Campbell called it "bliss" and advised his students to "Follow your bliss."

In terms of occupation, I usually ask people: What job or role in life can you imagine that would make every day sing? What would you pay somebody to let you do all day, every day?

This is a multi-dimensional exercise. As we present it here, with its focus on several areas of life rooted in NOW, it provides a realistic means for creating a motivating and empowering vision of an attainable future. Because it is rooted in NOW, the future that you build will be one that is close to your heart. But the exercise can also be used (starting with step one) as a means for testing the value and enhancing the motivational properties of any outcome or goal.

The exercise begins with the felt experience of a future rooted in NOW. It gradually awakens the details of that experience in very concrete terms—what you are doing. Once the vision is created, it goes through the process of clarification using the numbered steps. Much like the techniques in Exercise One, each step adds another layer of sensory and emotional experience so that the future becomes more and more real. The image becomes so real that it becomes very clear whether you want it or not. If you want it the last steps are designed to help you get there.

Instructions

Begin by firing off your resource state (NOW) and create an inner experience (one at a time) for each of the following categories. Do this by floating out into the future or by opening a window in your imagination and stepping out into the future.

As you perform the exercise, don't go to the future with a plan in mind. Let the plan or outcome emerge from the experience of NOW. In each category, experience what you are doing, not what you have.

1) Spiritual life. When you get in touch with who you really are, what kind of spiritual aspirations do you have? Who would you emulate? What would you do differently? How would you feel? How does this relate to or emerge from the resource state? Use NOW to experience your spiritual life in the future.

Notice how you give physical expression to this reality. Notice the kinds of things that you are doing, especially the things that increase and deepen your ability to grow as an individual in a spiritual context.

2) Relationships. When you fire off your resource state and think about your future relationships; what will you be doing differently? What will have changed in the way you respond to others and they, in turn, respond to you. How does this relate to or emerge from the resource state? Notice how you give physical expression to this reality. Notice the kinds of things that you are doing, especially the things that increase and deepen your ability to grow as an individual in relationship to others.

3) Intellectual life. Fire off your resource state and think about how you use your brain and your mental capacities. How are you using them differently? What will you be doing more? What will you be doing less? What challenges you and how do you enjoy it? What is your next level of educational need? How do you feel? How does this relate to or emerge from the resource state? Notice how you give physical expression to this reality. Notice the kinds of things that you are doing, especially the things that increase and deepen your ability to grow as an individual and in your ability to use your intellectual abilities and gifts.

4) Occupation/work life. Fire off your resource state and ask yourself what you must do (or do more frequently) that will allow you to feel this way more often. Do you have a different job? Have you started a new career direction? Do you work less? Have you started a hobby? What is different? What are you doing in this future and how does it make you feel?

5) Health. Fire off your resource state and ask yourself what you must do (or do more frequently) that will allow you to feel this way more often. Have you changed jobs? Have you changed eating or exercise patterns? Do you work less? Take more vacations? What is different? What are you doing in this future and how does it make you feel?

Use the following procedures to test and concretize their outcomes. You can talk them through these or provide a written list. If any of the steps is unclear or difficult to resolve, have them fire off NOW and allow the answers to emerge from within the state.

1. <u>State your experience of that future as a positively desired, outcome. Express it in terms of behaviors: what you will be doing.</u>

Think about what you have experienced in your vision of the future. Begin to describe it as a positive thing, something you can hold in your hand or put in a wheelbarrow or something you can do.

I will get my Associate's Degree is a good outcome. You can see yourself at graduation.

I will begin doing more of my art, my music, my…. I will spend more time using deep states to access new capacities. I will be more familiar with my internal landscape and will have new ways of generating options. I will have new choices in how to spend my time and how to focus my attention. I will finally be able to….

2. <u>Make sure that it is under your personal control.</u>

A proper goal must be under your control. It must be something that you can do: saving money to buy a house or business, getting the training and connections to make a career change, finding new ways to change the way you think or behave, leaving an old relationship or starting a new one. These are all things that you could do. All of these things are choices that would be under your control.

Under your control implies that it does not depend upon someone else or some external agency. It is something that flows from your own capacities. If it depends on someone else's action, it is probably a pipe dream.

As you focus on the NOW state and how you will be feeling after using it for five years and growing with it, what new sense of your own capacities arise? What do you know that you can do, because you are feeling more like this every day?

Personal control also includes reasonableness. Is the goal you want realistic, or should it be broken up into sub goals or outcomes?

3. <u>Specify three different ways in which you will know that you've gotten it when you get it.</u>

How will you look when you have it? How will you feel? Who will be there? What will you see and feel? The more fully you can imagine getting it or doing it, the more powerfully motivated you will be to get it. Use all of your senses.

Fire off NOW and imagine that you are there and you can see it and feel it and touch it. It is really important in this step to feel and see and hear and taste and smell your future reality. The more senses you can use the more motivating the goal will become. The motivational centers of the brain encode preference and motivation in terms of full, multisensory representations.

If you're thinking about a restaurant business, envision the deed or licenses, hold them in your hand, feel yourself signing the deed, writing the menu, opening the front door on the first day of business. Smell the food cooking in your ovens.

Feel the handshake from the boss and see his face as you are given the promotion.

Smell the fresh ink on that new book. Feel its heft in your hand and the crispness of the pages. See your name on the cover and on the spine.

See the body language and hear the tones that tell you that that relationship really has changed. Notice the how you hold yourself and the people that you love. Notice thir responses to the new you.

Feel the exhilaration as you cross that finish line or reach that peak of physical activity. Notice the positive change in your muscle tone, your energy, your breathing. Notice the change in your gait and posture. Discover new kinds of personal satisfaction.

See the changes in your face, posture and movements as you live out new spiritual realities. Become aware of the depths and directions of feeling and experience as you awaken into that new reality.

For spiritual outcomes, fire off now and notice where that leads you spiritually. What kinds of practices behaviors or realities open to you and support this direction of personal growth? How do the future experiences impact the depth and richness of your current experience? As you focus on the practice of using the anchors, how does your experience--after five years--feed back into the way they work now?

For this step, specify three of these sensory experiences that will really let you know when it's yours.

4. <u>Decide where and when it will be appropriate. Will you want this all the time? Is it appropriate everywhere? Should it be limited to a specific context?</u>

As we make a goal realistic, it is important to understand that it may not be appropriate at all times and places.

Fire off NOW and consider where this fits in your broader circumstances. How does this outcome fit in your life priorities from the perspective of NOW? What kind of priority does it have and when will it be good to work on it? From the perspective of NOW, are there other things that are more important or pressing? How does this integrate or resonate with the other changes you are planning?

5. <u>What will it change in your life and in the lives of the people around you?</u>

Real goals have real consequences. When you are sober, you will have new friends and new relationships. How will this work for you? Who will support you? Who will resist you? What will it cost in terms of relationships? What will it get you? Are you willing to take the chance? What will you gain in terms of new opportunities?

When you enter a new business or a new neighborhood, there are also changes in your life. How many hours will you be working now? How many nights, how many weekends? From the perspective of now, begin to experience with a sense of appreciation, the kinds of changes that you will be experiencing in the world around you and notice how well your new perspective prepares you for them.

New spiritual insights often require new ways of being in the world and managing your time. How will these things change? New relationships and changes in old ones make specific demands on time, money and other friendships. How will these things change for you?

6. <u>Experience now, in your imagination, how you will look and feel, what you will see and hear when this is a reality.</u>

Use NOW to go back and get the image again. See your surroundings in bright color. Experience the people around you. Experience in full sensory detail, how you will feel and how those around you will respond. Step all the way into your own body and experience the future from inside.

What we are accomplishing here is getting in touch with your future self; the self who has already accomplished your goals. It is important to feel and identify with this future you because he or she will show you how to get where you want to be.

Use NOW to fill out the details. With each pump, begin to notice how more and more detail arises into consciousness. Go through all of your senses. Step all the way into it. See it, feel it and hear it from your own perspective. How do you feel having it? How do you hold yourself? Move into that same posture. What do you say to yourself? What do the people around you say?

7. <u>Move backwards from the final realization of the goal to discover the steps that make it possible.</u>

Once you have a real experience of yourself having what you want, take time to get into it. Enjoy it. Feel it and continue to feel it. Then, from that place, where you can see it and feel it and taste it and hear it, look back towards today and find the steps that got you there. Begin with the last step, the finishing touch. Really be there. Ask yourself "What was the last step that I made that put it together?" Take your time. When you've answered that question, ask: "What was the step before that? ... and before that?" Take the time you need to find the steps that got you there. Keep the state, feeling like you've already gotten it. This will guide your mind to the steps that got you there.

Make sure that there is a manageable number of steps and not too many. Ask your unconscious to help you to make the size of the steps just right.

8. <u>List the five steps necessary to get from here to there.</u>

Think about the steps that you just learned from the future you. Break the list into five steps that you can handle. If necessary, the five steps can be five sub goals and you can use the process on each of the sub goals.

Outcome Worksheet
OUTCOME :

1. Is it stated in the positive, or can it be stated in the positive? State it.

2. Is it under your personal control? How?

3. Can you specify three different ways in which you will know that you've gotten it if you get it?
   A)
   B)
   C)

4. Do you want this all the time? Is it appropriate everywhere? Should it be limited to a specific context?

   When do you want it?

   When don't you want it?

   When is it right?

   When is it wrong?

5. What will it change in your life and in the lives of the people around you?
   Be specific:

6. Experience now, in your imagination, how you will look and feel, what you will see and hear when this is a reality.
   Describe what you see and hear and feel. Who is there? What is it like? Be there Now.

7. Move backwards from the final realization of the goal to discover the steps that make it possible.
   List the steps

8. Enumerate five steps necessary to get from here to there.
   1).
   2).
   3).
   4).
   5).

Further Applications
Setting goals is an important task in modern life. What are other broad areas of your life that you would like to explore? When outcomes are rooted in NOW, they produce powerfully motivating futures. The exercise can also be used to evaluate current outcomes. Use the Outcome Worksheet to test your current goals. If they are truly yours, they will become even more desirable. If they are not yours, or just plain wrong, you will really become aware of that too.

References for the exercise

Bandura, A. (1997). *Self-Efficacy: The exercise of control.* NY: Freeman.

Baumeister, R. F., & Heatherton, T. F. (1996). Self regulation failure: An overview. *Psychological Inquiry,* 7(1), 1-15.

Bodenhammer, B. G, & Hall, L. M. (1998). *The user's manual for the brain: The complete manual for neuro-linguistic programming practitioner certification.* Institute of Neuro Semantics.

Cade, B., & O'Hanlon, W. H. (1993). *A brief guide to brief therapy.* NY: W.W. Norton.

Csikszentmihalyi, M. (1990). *Flow: The psychology of optimal experience.* NY: Harper and Row.

Damasio, A. R. (1999). *The feeling of what happens: Body and emotion in the making of consciousness.* NY: Harcourt.

Deci, E. L. & Ryan, R. M. (2008). Facilitating optimal motivation and psychological well-being across life's domains. *Canadian Psychology* 49(1), 14–23.

Dilts, R., Delozier, J., A., & Delozier, J. (2000). *encyclopedia of systemic neuro-linguistic programming and nlp new coding.* Scotts Valley, CA: NLP University Press.

Dilts, R., Delozier, J., Bandler, R., & Grinder, J. (1980). *NLP, vol.1.* Capitola, CA: Meta Publications.

Erickson, M. H. (1954). Pseudo-Orientation in time as an hypno-therapeutic procedure. *Journal of Clinical Experimental Hypnosis,* 2 ,261-283. In Milton Erickson and E. L. Rossi (Ed.) *The Collected papers of Milton H. Erickson on hypnosis: Vol. IV. Innovative hypnotherapy.* NY: Irvington. 1980.

Gray, R. M. (1997a). Ericksonian approaches to the ego-self axis: Establishing futurity and a sense of self in addictive clients. Seminar: *Innovative Approaches to the Treatment of Substance Abuse for the Twenty First Century.* St. Francis College, Brooklyn, NY. Published on the WWW at http://www.temperance.com/nlp-addict/articles.html

Gray, R. M. (2001). Addictions and the Self: A Self-Enhancement model for drug treatment in the criminal justice system. *The Journal of Social Work Practice in the Addictions, 2(1).*

Gray, R. M. (2002). The Brooklyn Program: Innovative approaches to substance abuse treatment. *Federal Probation Quarterly. 66(3),* December 2002.

Gray, R. M. (2005). *Thinking about drugs and addiction.* Boulder CO: NLP Comprehensive. http://www.nlpco.com/articles/AddictionsGray.html

Gray, R. M. (2006). About addictions: Notes from psychology, neuroscience and NLP. Raleigh, NC: Lulu Press. http://www.Raleigh, NC: Lulu Press./content/3497961

Grinder, J. & Delozier, J. (1987). *Turtles all the way down: Prerequisites to personal genius.* Scotts Valley, CA: Grinder and Associates.

Iacoboni M., Woods, R. P., Brass, M., et al. (1999). Cortical mechanisms of human imitation. *Science. 286*: 2526-2528. Reported at

Koestner, R. (2008). Reaching one's personal goals: A motivational perspective focused on autonomy. *Canadian Psychology, 49*(1), 60-67.

Laski, M. (1961). *Ecstasy in secular and religious experiences.* NY: Jeremy Tarcher.

Linden, A., & Perutz, K. (1998). *Mindworks: NLP tools for building a better life*. NY: Berkley Publishing Group.

Maslow, A. H. (1943). A theory of human motivation, *Psychological Review, 50*:370-96

Miller, S. D. & Berg, I. K. (1995). *The miracle method: A radically new approach to problem drinking.* NY: Norton.

Notz, W. W. (1975). Work motivation and the negative effects of extrinsic rewards. *American Psychologist* (September 1975), 884-891

Ostrander, S., & Schroeder, L., with Ostrander, N. (1994). *Superlearning 2000.* NY: Delacorte Press.

Overdurf, J. (2006, April). *You never know how far a change will go …Beyond goals.* Pre-Conference workshop conducted at the 19th Annual Convention of the Canadian Association of NLP. Retrieved on April 15, 2008 retrieved from http://johnoverdurf.typepad.com/ canlp/files/ canlpmanual.pdf

Robbins, A. (1986). *Unlimited power.* NY: Fawcett Columbine.

Wegner, D. M., Schneider, D. J., Carter, S., & White, T. (1987). Paradoxical effects of thought suppression. *Journal of Personality and Social Psychology, 53*, 5-13.

Setting Goals that Work is copyright, 1993 by Cunningham and Gray Resources.

Exercise 11

ImageStreaming Into The Future

Presuppositions Underlying the Exercise:

ImageStreaming is a technique developed by educator and futurist, Win Wenger, Ph.D. (1993). ImageStreaming is rooted in his analysis of the relationship between measured intelligence and the capacity of highly intelligent and creative people to make links between visual information and verbal labels. More importantly for our purposes, ImageStreaming helps to articulate the outcome and links it to a positive and empowering subjective experience.

The ImageStreaming exercise assumes a high degree of subjective sensory awareness. This is assumed to have developed for all participants who have conscientiously applied themselves to the Program. Persons who complain of their inability to visualize are suspected of having been non-compliant through the other exercises.

Throughout the exercise, the idea of visualization is repeated. While at times this may in fact become eidetic or dreamlike, for the most part it will remain somewhat fuzzy and daydream-like. Both levels are acceptable.

The exercise presupposes the validity of Wenger's technique for engaging unconscious problem solving and creative capacities through the use of these techniques. It also presupposes the value of declaring and describing one's plans aloud. This follows not only the findings of Cialdini (1993), but is further recommended by Prochaska, Norcross and DiClementi (1994) as an important part of movement towards the action stage of change.

Expected Outcome

Participants will express a clearer sense of the process of achieving the goals stated in the last exercise. They will be able to express a sense of their increased reality. Participants who complete the exercise find it very enjoyable.

Instructional Notes

Introduce the ImageStreaming portion of the exercise by describing the process and its relationship to increased intelligence and creativity. Inform the participants that it is based on an educator's analysis of Einstein's thought patterns and how he came up with his great ideas. You may further wish to inform them that it has been found that people who practice ImageStreaming report increased creativity.

Have the group divide into pairs. One person works as the goal seeker and the other as the operator or partner. The partner is responsible for asking questions and ensuring that the goal seeker meets the exercise's criteria at each stage. S/he ensures that all of the answers are appropriately concrete and that each criterion is met in turn. They should go through the entire exercise using the worksheet provided.

When they reach the ImageStreaming step at stage six, the goal seeker must access the goal state in his or her imagination and, without hesitating or self-editing, talking as fast as he can, describe the visual, kinesthetic, auditory and olfactory elements of the outcome situation aloud to the partner. It is the partner's responsibility to encourage the goal seeker, to ensure that the process proceeds quickly and to provide continued prodding for the production of imagery. At this stage of the exercise, participants should be expected to take a minimum of ten minutes each to ImageStream the details of their outcome. It must be emphasized that the process must be done as fast as possible—describe each image or sensation as it comes— and it must be done without self-editing.

If the process slows, the partner may prompt the goal seeker by asking sensory-based questions. What color is it? Where are you? Who is there? What are they wearing? What are you wearing? What are the smells? Are there sounds? What kinds of sounds are they? Describe their qualities. The partner must be demanding and should often prompt the goal seeker to speak faster and with more detail.

It is never the partner's task to edit or comment on the images themselves. One of the important characteristics of the exercise is that it is unedited. The partner's task is to encourage the fast, free flow of images. It is not a conversation.

If the goal seeker gets really stuck, (i.e., they can't seem to come up with anything else to describe) the partner may suggest to them that they imagine a door that leads into an unknown space. Behind that door is an important aspect of their outcome which they have not yet considered. Have the participant describe the door in great detail, using every sense. Once the door has been imagined and fully described; at the partner's command, let the goal seeker open the door and begin to describe whatever comes to mind. The goal seekers should be reminded to allow themselves to regard the door with a great deal of curiosity and to allow themselves to be surprised by what they find on the other side of it.

Some ImageStreams will very concretely remain with the details of the imagined future. Others will drift off into dream images and phantasmagoria. These are both acceptable. ImageStreaming seems to tap into much the same capacities as dreaming does when it becomes a solution-finding experience. Let the images continue for as long as they will with the goal seeker enjoying them and describing them to the partner.

At the end of the ImageStreaming segment the goal seeker will once again record a set of steps for getting to the imagined outcome from the perspective of being there. If the ImageStream segment departs dramatically from the actual theme of the outcome, have the participant reestablish their conscious, present-tense connection to the future outcome before proceeding.

In all cases, do not allow the exercise to become conversational. The seeker is describing internal imagery; the partner is guiding the process.

Behavioral Standards

Participants who complete this exercise appropriately emerge with a clear and expressible sense of their own connection to the outcomes or goals with which they worked. Strong expressions of Self-Efficacy and descriptions of creative leaps are common.

Meditation

Any meditation may be used. It may be appropriate to leave the choice up to the participants.

Exercise 11
Image Streaming Into the Future

ImageStreaming is a technique developed by Win Wenger, Ph.D. to allow other people to use Albert Einstein's creative strategies. According to Einstein, most of his discoveries were made as visual thought experiments that resulted in a felt sense of the relationships involved. Einstein reported that his discoveries were not mathematical in nature, mathematics were only the language that he used to express them.

In the literature of creativity, the most important task is not the flash of insight, but the tedious work of bringing the insight into expression. For Einstein, this meant finding a way to express the felt intuitions that were the result of his thought experiments.

Wenger suggests that this process, which he calls pole-bridging, involves finding new ways to make connections between the brain's rich store of sensory information and the relatively small verbal repertoire that most of us possess. His technique, called ImageStreaming, involves learning to pay attention to and to describe the spontaneous sensory output that emerges from the unconscious as we work on a project or seek to answer a problem. By linking the flow of private, non-verbal information to our conscious verbal capacities, we can increase our own creativity.

In many ways this technique makes conscious use of the kinds of process that appears in dreams. In dreams, ideas are connected by their emotional relevance and answers worked out through symbols.

In the current exercise, we will be using ImageStreaming to make the felt reality of the futures that you have visited eve more real. We will also invoke the capacity of the unconscious to add details to the vision in a way that is consistent with the NOW state.

ImageStreaming into the Future

For this exercise, you will need a partner or a tape recorder. If you use a partner, you and your partner will take turns going through these steps with each other.

There are a few basic rules for ImageStreaming. 1. Speak out loud to someone or something. Just thinking about it doesn't have the same feel or impact. 2. Talk as fast as you can without self-editing. It doesn't matter if it's foolish, bad, obscene or crazy; just keep talking. 3. All of your speech should be in sensory based language. Talk about what you see and hear and feel. Talk about the colors and the sounds and the textures that appear in your mind. Speak in the present tense. Talk about what you are experiencing NOW. 4. If you get stuck, go back to the original image or question and start over.

In the first steps of the exercise we will review the outcome rules from the last exercise. The actual ImageStreaming begins at step six. Take about 15 minutes for each participant.

1. Goal Seeker: Get a partner and state your outcome.

    Partner: Ensure that the outcome is stated in the positive. If it is not or cannot be restated, choose another outcome.

2. Goal Seeker: Tell your partner how the goal is under your personal control.

Partner: Make sure that you understand that this is something that they can do and that they are personally responsible for doing it.

3. Goal Seeker: Tell your partner three different ways in which you will know that you've gotten your outcome. Tell your partner in as much sensory detail as possible how you will know.

    Partner: Ask questions like: How does that feel? What does that look like? What do you smell? Ask sensory based questions in the present tense.

4. Partner: Ask the goal seeker the following questions: Is it a full time effort or are there other things that you must care for? Do you want this all the time? Is it appropriate everywhere? Should it be limited to a specific context? When do you want it? When don't you want it? When is it right? When is it wrong?

    Goal Seeker: Be as specific as you can about when you will be doing this or working on this.

5. Partner: Ask the goal seeker the following questions: What will it change in your life and in the lives of the people around you? Relationships? Schedules? Habits? Recreation? Family life?

    Goal seeker: Be as specific as you can about what will change in your life:

6. Goal Seeker: Close your eyes and fire off NOW. As NOW grows in strength, experience in your imagination, how you will look and feel, what you will see and hear when this is a reality. Go into your imagined future and describe to your partner in as much detail as possible what you see and hear and feel. Who is there? What is it like? Be there now. Do it fast, do it out loud. If you run out of ideas, start over. If the image changes from the original outcome, just describe what you experience.

    Partner: Continue to prod the goal seeker for details. Do not get into a conversation. Demand colors and shapes and sounds and smells and textures. Keep the descriptions flowing. Do not judge any expression. Do not allow the Goal seeker to judge.

7. Goal Seeker: Staying in the future state, continuing to feel good, look back from that future towards the day of the exercise. Describe the steps that got you there, beginning with the one closest to the goal. From there move backwards from the final realization of the goal to discover the steps that make it possible. List the steps.

    Partner: Record these steps.

    Each step must be a concrete action. If you can't act it out, hold it in your hand or put it in a wheelbarrow, it is not concrete enough. Go back and get some real answers from your future self.

8. What must you do now and why is it important?

    What are the other steps?

    Break the list into five steps that you can handle. If necessary, the five steps can be five sub goals and you can work the process on each of the sub goals.

    List the five steps necessary to get from here to there.

Further Applications

The power of telling someone, of speaking aloud, is often underestimated. Notice the difference between talking to someone or even speaking into a tape recorder and just thinking. Feedback, hearing yourself make the commitment, seems to be part of it. The other part is the kind of mind set and felt experience that declarative speech has.

Try ImageStreaming outside of the context of the exercise. Get a partner or a tape recorder. Ask your unconscious mind a question and begin to describe the images as fast as you can. Do it for 10 or 20 minutes. When you are done discuss the common themes and images with your partner or write them down. Notice what new ideas come to mind.

References for this exercise

Cialdini, R. B. (1993). *Influence: The psychology of persuasion.* NY: Quill Publishers.

Dilts, R. (1997). *Strategies of genius (Vol. 2).* Cupertino CA.: Meta Publications.

Jacobi, J. (1974). *Complex, archetype, symbol in the psychology of C. G. Jung.* Princeton: Princeton Univ. Press.

Wenger, W., & Poe, R. (1997). *The einstein factor.* Rocklin, CA: Prima Publishing.

Exercise 12

Sponsoring a Potential

Presuppositions Underlying the Exercise:

Sponsoring a potential is taken directly from the work of Robert Dilts, one of the originators of NLP and one of its most well-known exponents (Dilts, 1998). The exercise works on the same principles of Pseudo-Orientation in Time, as do the Smart Outcome Exercises and the Future Perfect Meditation. In this exercise there is a further presupposition of the need for some outside person to take responsibility to nurture those talents and abilities that are needed in order to reach the anticipated future. It is the only exercise in which we posit some deficit. Nevertheless in positing the deficit, we also provide a present-state, experienced resolution.

The assumed deficit lies in the failure of the environment or some external figure to nurture some specific personal capacities that might have led to a very different outliving of this individual's capacities. As the individuals in this program have often experienced emotional, social and economic deprivations, there is no end to the kinds of problems that can arise here. Nevertheless, because this has been a skills-oriented program, most participants are able to focus on some one need that might have made all the difference in the world.

The outworking of the exercise presupposes that most people organize their time in linear fashion on a metaphorical time line (Dilts, 1993; James & Woodsmall, 1988). For most (normally organized, right handed Westerners) people, this means that time is represented with the past somewhere behind and to the left and the future is somewhere out in front and to the right. The use of a metaphorical time line can be used to change our current-time experience of past experiences and to shape future realities.

In this exercise the participants have the experience of setting up their own time line. As they walk through the time line, re-experiencing the major events of their life, they are accompanied by the Facilitator or another participant who, acting as sponsor of the potential, acknowledges, protects and symbolically nurtures that skill or potential.

There is a presupposition on the part of the participants that they can understand and perceive the participant's capacity to attain this goal or have this quality and that they can congruently express that perception.

In the assumption of the role of sponsor, the exercise assumes that the participant or Facilitator will focus their attention on the best interests of the client and will further project their good intentions in a congruent and acceptable manner.

Expected Outcome

Participants will establish a metaphorical time line. They will identify a potential or skill that was never nurtured that would have made a significant difference in their life. They will have an experience of revisiting their life and the major events in it as if it had been done with the support they needed to realize the target potential. They will communicate with their future self and receive a message of encouragement from the future.

Instructional Notes

This exercise is very powerful. One of the key marks of real cooperation is the solemnity and impact that it has on the participants. It is often described as feeling like a marriage or initiation. As a result, we have saved it for the last formal exercise.

There are several important practical matters in the current exercise. For small groups the Facilitator can perform the role of sponsor for each of the participants. For larger groups it can be useful to have the group split up into groups of two or more, and have other members perform the roles of sponsor and client.

The first part of the exercise requires the client to mark out a time line on the floor or on the ground. The client should orient the time line in a manner that is comfortable for them. On the time line they need to mark a place for the time of their own conception, the chronological present and a future when their potentials will have been largely realized.

The exercise begins with the client standing facing the time line looking towards the future from a point (off the time line), just before conception. The sponsor stands to the right of the client (also off the time line), facing her.

The core or essence to which Dilts refers as the beginning point is expressed very well in our complex Anchored state associated with the word NOW. It has been our intention all along to use this state to constellate a sense of deep Self as center, potential and direction.

In this exercise, the participants should use the NOW Anchor to reach inside for a sense of who they are now and have always been. As they explore this deep sense of Self, they will need to identify some part of themselves that was never properly developed. This is, a part which, if it had developed, would have resulted in a significant change in the quality of their life.

This missing element should be one which "If it had been nurtured or protected during your life, it would have allowed you to more fully: (a) establish healthy boundaries, (b) overcome barriers, or (c) evolve yourself more completely" (Dilts & Delozier, 2000).

Healthy boundaries reflect the way we deal with other people in our lives. If boundaries are too fluid, we lose ourselves in others and fail to establish a clear sense of Self. Boundaries that are too rigid result in isolation and exclusion. In order to preserve Self, the individual blocks itself off from genuine, caring or deep relationships. Overcoming barriers suggests that sometimes there is some part that would have energized our existence. Had it been there, nothing would have stood in our way. Evolving one's Self more completely suggests that we are in control of what we are becoming. What part of your life that went unprotected and undernourished would have changed your potentials in a significant manner?

Once the appropriate potential or trait has been identified, the client is to create a symbol for it.

When the participant has chosen a sponsor, or the Facilitator comes to take the sponsorship role, the client must now share their symbol and the resource that it represents with the sponsor. The sponsor may need to ask questions until: 1) S/he fully comprehends the nature of the potential or resource. 2) S/he can see its importance to the client and 3) S/he can genuinely see that this really is a potential possessed by the client.

When done as a group exercise, it is useful to have all of the participants, who are not otherwise engaged, fire off NOW, and visualize the client's symbol manifesting in his heart.

When the sponsor can really "see" that this potential resides within the client, he offers her his right hand. The client takes the hand and places it with her hands over it, above her heart. (Please note that the heart area, a few inches below the neck, can be touched without invading more private areas.) In response, the sponsor places his left hand in the center of her back. In this position, with her heart shielded by the hand of the sponsor and her back supported by the sponsor's other hand, the client steps onto the time line. As the client walks down the time line, both should be focused on the potential. All group members present should likewise enter the NOW state and project their positive assent and support towards the client.

The client may be reliving significant life events as they pass them on the time line. They are relived, however, with the difference that now there is a nurturing hand supporting and protecting the resource through these crucial passages. Let the client note the changes in her experience as they are relived with the aid of the sponsor's physical support.

If there have been many significant passages, the client may choose to pause at the more important ones to experience the difference that this support makes. The sponsor should be careful to match the client's speed, keeping his attention on the client and the resource/potential. Take as much time as necessary to pass from conception to the present.

When you reach the place marked out on the time line as the present, the client should take as much time as necessary to internalize the feeling given by the sponsor and the Group. When the client knows that he can proceed without the physical support of the sponsor, let him remove the sponsor's hand from his chest and replace it with his own hand as an anchor in the same place. The sponsor should simultaneously remove her hand from the client's back. Now, holding his own hand over his heart and so, anchoring the sense of support, the client walks at his own speed from the present into the future. During this part of the walk, the sponsor walks alongside with arms extended but not touching client. He now provides a supportive environment but no physical anchor.

When the client reaches the future, let him pause for a minute. Get a sense of how it feels and then turn around and look down the length of the time line. Get a sense of how far it is.

Sponsor and client then walk off the time-line and back to the moment before conception. The client again gets into contact with his core and the symbol of the potential. Simultaneously, the group should refocus on NOW while continuing to project their positive intent towards the client. Now the client walks through the entire time line with his hands firmly anchoring the spot over his heart. He is accompanied by the sponsor but they do not touch. The client takes as much time as necessary to pass through all of the important places between conception and the present (as marked on the time line). When he reaches the future, he turns around and looks at his own (imagined) figure standing on the place marked as the present. From that place in the future, he sends a message of hope and encouragement to himself in the present time. The message should be addressed to the present-time image in second person —"You." The client then walks to the place marked as the present, turns to face the future and imagines seeing his future self, and hearing his message. Share the message with the sponsor and the group (if appropriate). Debrief.

Behavioral Standards

At the end of the exercise participants will have a powerful sense of their own positive futurity. Many will experience a restoration of missing or unfulfilled potentials. Others will have near epiphanies. For many it will be a deeply moving experience. Disassociated or unconnected responses are suspicious.

Meditation

No meditation is usually necessary after this session

Exercise 12
Sponsoring a Potential

The following is an example of a sponsorship exercise, developed by Robert Dilts. It begins with the idea that people can take on the role of sponsor, a true witness and protector for the talents in themselves and others. Sponsors see the potential in others and provide a safe place for that potential to develop.

The exercise begins with the identification of an important potential that has always been true but for one reason or another, it was never honored, nurtured or developed. It is also a capacity which, in hindsight, would have made a significant difference in how you live, how you are developing and how you relate to others. It has not been fully developed because it has lacked the necessary sponsorship.

Sponsorship is provided by having a partner who is able to congruently "see" that potential. They will accompany you along your time line, through various key events of your life, safeguarding that potential and encouraging it to evolve.

A time line is the way we imagine time. Think about how we talk about the past as being behind us and the future before us. For this exercise, we will imagine that your time line can be projected on the floor or on the ground as a line. The line begins with a place that marks your conception at one end, the present somewhere in the middle and the future, at the other end.

If you are doing this outside of a group situation, have someone you trust act as sponsor. In a group, one person, the facilitator or the entire group can act as the sponsor.

The exercise involves the following steps:

1. Create a physical time line. That is, mark out on the floor or ground a line that you can walk along with points identifying conception, the present and the future when this potential has been realized.

2. Go back to the location that represents a point before your conception. Use NOW to get in touch with your core or essence. Your core consists of everything that was true of you then, has always been true of you and always will be true of you. It was there at the moment of your birth and it is still the same now.

3. Choose a characteristic or potential, that has always been within you, but it was never encouraged or protected. If it had been nurtured or protected during your life, it would have allowed you to more fully: (a) establish healthy boundaries, (b) overcome barriers, or (c) evolve yourself more completely. Create a symbol, or icon, for this resource or characteristic.

4. Have a partner be your 'Sponsor' or 'Guardian Angel'. Share the potential with your 'sponsor'. The sponsor is to listen and, if necessary, ask a few questions, until he or she is able to authentically and congruently "see" or sense the reality of this potential in you. Share your symbol with the sponsor.

5. Both you and the sponsor are now to get in touch with your own "cores" — the

parts of yourselves that you want to be, and believe you will always be; the deepest, most genuine part of yourself. When the sponsor is in touch with their core (NOW), and they are able to congruently sense your potential, he or she will signal you by offering you one of his or her hands. When you are ready to accept their sponsorship, signal your partner by placing their hand over your heart. The sponsor will then place their other hand on the upper center of your back.

6.   Focus on the resource or characteristic that you would like to have nurtured and protected and walk along your time line towards the present. As you do so, re-experience the important events of your life. Notice how the sponsorship of your partner helps to enrich and strengthen that resource/characteristic.

7.   When you reach the present, pause until you can really sense that you no longer require the sponsorship of your partner. When you know this, release their hand from your heart. The sponsor will then remove the other hand from your back. You may then continue walking to your own future. As you walk into the future allow the resource/characteristic to continue to blossom and develop. As you continue to walk into the future, your sponsor will walk along beside you. When you get to the future; look back down your time line. Get a sense of the distance between then (that future) and now (the place you marked out to represent the present).

8.   Return to the beginning of your time line and repeat the process, being your own sponsor. Fire off NOW. Get in touch with your potential, place your own hand over your heart, and walk up your time line. In this walk you will be accompanied by your sponsor (who will walk with you but does not touch you). Move all the way into your future. When you reach the future, look back towards the present (the place you marked out as the present on the time line). From the future end of the time line, imagine seeing yourself there, in the present and, from the future, send a message of encouragement back to yourself in the present. Then, return to the present (the spot you marked on the timeline), stand there looking towards the future. Imagine seeing the future you and hearing their message. Notice your response and share your experiences with your sponsor

Sponsoring a Potential is Copyright, 1998, by Robert Dilts, Unified Field Theory for NLP - Page 23, and is used with permission.

References for the exercise
Dilts, R. (1998). *Unified field theory of NLP*. Scotts Valley, CA: Robert Dilts.

Dilts, R. & Delozier, J. (2000). *Encyclopedia of systemic neuro-linguistic programming and nlp new coding*. Scotts Valley, CA: NLP University Press.

Gray, R. M. (1997b). Addiction and the nature of meaning: Reframing in substance abuse treatment. Seminar: *Innovative Approaches to the Treatment of Substance Abuse for the Twenty First Century*. St. Francis

College, Brooklyn, NY. Published on the WWW at http://www.temperance.com/nlp-addict/articles.html

Gray, R. M. (2001). Addictions and the Self: A self-enhancement model for drug treatment in the criminal justice system. *The Journal of Social Work Practice in the Addictions, 2(1)*.

Scripts

The following scripts are provided to guide you with the basic language needed to create and intensify the basic states and anchors. They are guidelines and need not be followed slavishly. Some of the texts have been marked so as to give you an idea of how to break-up the phrasing so as to maximize their impact.

Script One Enhancing resource states

SCRIPT ONE   Exercises One and Two and as Needed for Review.

Use this script for the first several passes through exercise one. It may also be used anytime that clients are unsure of the process or unable to create the altered states.

1. **Invite the participants to choose an experience that made them feel wonderful.** It may have been empowering, fulfilling, fun or ecstatic. Let them keep it private. The aim of the first exercise is to gain experience with the techniques.

2. <u>In the first exercise</u> *we do not specify the state* and do not ask the participants to describe the state, unless they have troubles later in the exercise. *This means that some of our participants will use illegal, immoral and otherwise objectionable states. This is not a problem.*

    <u>For the second exercise</u> Invite the participants to begin with one of the exemplars (Focus, Solid, Good, Fun or Yes). In this exercise all participants are asked to access the same state.

3. **Have the participants close their eyes and experience the memory. Let them note just how they get to the memory: what they notice first, a picture, a smell, a feeling? What comes next and next and next?**

4. **As they access the states, they are asked to notice the difference between associated and dissociated experience—in the picture or out of the picture. They vary the intensity, — bring it closer, make it brighter, make it louder. After each change they are asked to note the change in their felt experience. Each instruction is designed to provide a felt change in the experience and to provide practice in the manipulation of feeling by changing the submodality qualities of the experience.**

5. **During the initial walk-throughs, as you are reading the submodality lists, make an effort to use normal intonation and volume, with no effort at trance language or special emphases.** Persons in substance use treatment are notoriously paranoid and can respond badly to unconscious communication styles if used prematurely. Use trance patterns only after they have created altered states for themselves. These will be treated later in the training.

    - People differ as to which sense arises first when they access a memory. Some people remember pictures; some sounds; some begin with feelings. Most people in the West prefer vision. For this reason, we are starting with the visual part of the memory. If you find that sound or feeling comes up first for you, feel free to start there and return to the other senses in the way that works best for you. But please read through the rest of the exercise before starting.

    - One more thing; All of this is easy. Most of it consists in just noticing how things are. The simple act of turning your attention to the sensory distinction is often enough to change it. In every other case, gentle imagining works fine.

<u>Script for enhancing resource states</u>

1. <u>For Exercises Two and later.</u> Invite the participants to begin with one of the exemplars (Focus, Solid, Good, Fun or Yes).

<u>For Exercise One</u> have them use the same state that they have been working on, or have them choose a new one after discussion.

2.   As they access the states, they are asked to notice the difference between associated and dissociated experience–in the picture or out of the picture. They vary the intensity, — bring it closer, make it brighter, make it louder. After each change they are asked to note the change in their felt experience. Each instruction is designed to provide a felt change in the experience and to provide practice in the manipulation of feeling by changing the submodality qualities of the experience.

3. Have the participants close their eyes and experience the memory. Let them note just how they get to the memory: what they notice first, a picture, a smell, a feeling? What comes next and next and next?

4. Notice whether, in your imagination, you are experiencing the memory from within, or watching it from outside like in a movie.

5. If your memory seems to be just in your head, imagine that you can *step all of the way into it*. As you experience the memory, you may even notice flashes that feel like really being there, focus on these. Take a few minutes to make sure that you are actually in the experience. Once you have the sense of really being there, even if it was only for flashes, come fully back into the present context.

6. Once you have a sense of what it's like to relive the memory from within, step all the way into it and get a feel for it. Notice that you can step right into one of those parts where it all came alive. Step right into it. Notice what you are seeing and feeling and hearing. Notice the patterns of tension in your muscles. Notice who is there and how you feel emotionally. Take a few minutes to get really familiar with the feel of being there. Enjoy it. Come fully back into the present.

<u>From this point forward drop the use of the word memory and begin to refer to *the feeling, the experience or the resource.*</u>

      Are you experiencing the resource from within your own body or watching it from outside, like a movie? If you are watching it like a movie or a video, step all of the way into the picture and experience it from within your own body. See what you saw, hear what you heard and feel what you felt from within your own body.

       Once the client has stepped into the experience, they can then begin to vary the submodality structure of the memory. Instruct them to make the changes in a way that makes the experience work best for them. Let them experiment with each dimension to find a level that feels best.
       Go through these one at a time, pause after each to allow for their processing time and, in the early exercises, ask them to describe how each change affects the experience. Tell them to remember the ones that work best.
       After each change, ask them to note the change in their felt experience. Each instruction is designed to provide a felt change in the experience and to provide practice in the manipulation of feeling by changing the submodality qualities of the experience. Remind them to take note of the kinds of perceptual changes that make the most positive difference in the experience.
       Imagine that you can bring the picture closer and make it bigger and brighter. Each of you has had the experience of changing the size of a picture on a computer screen and changing the brightness on a television set. Use your imagination, now to make the image closer, bigger and brighter. Notice how that changes your experience. Keep the change that produces the best experience. If the experience lessens put the experience back the way it was.
       Notice whether the image is in color or black and white—if it is black and white, turn on the color. If it is already in color, turn up the intensity. Notice how that changes the experience. Find out how much color feels best.

> Notice whether the experience is moving or still. If it is still, turn on the motion. If it is moving, turn it into a still image. Notice which one feels best. Keep the change that makes the experience feel best.
>
> Notice whether the sound is on or off. If the sound is off, turn it on. Adjust the volume of the sound so that it enhances the experience.
> Notice how the sound moves. Notice how the sound moves with the experience. Notice whether the sounds are noises or music or voices. Pay attention to where they come from.
> Breathe in through your nose and smell the smells that were present there.
>
> Come back for a moment, shake out the experience and talk to me about what happened. Did that feel good? Did you know that you could do that?
> What worked best for you?

After a few minutes of discussion invite them to just close their eyes and return to the place where they left off and continue as follows.

> Now, step all the way beck to the point where you just left off. For some of you the memory has gone away and you were just out there floating, that's good, go back there. Go back to the state where you left off and notice how easy this is.
> Notice how you breathe in this experience. Notice how you hold your body --the patterns of tension and relaxation that enhance your experience. Adjust your posture, so that it enhances the experience.
> Notice how you breathe in this experience and the expression on your face. Adjust your expression so that it enhances your experience.

7. Return to the experience and zoom right back to the very best part. Turn up the brightness, bring it closer and turn up the volume on the sound. While you do these things, note the path of the energy through your body. As you notice the feeling getting stronger, loop the feeling back through the starting point so that it doubles up as it moves through you. Notice that it moves further, faster and more powerfully.

8. Again, have them come fully into the present and shake out the state.

9. **Now have them return to the state and quickly enter into the feeling state.** As they note the rush of onset (call it a "rush") as the experience reasserts itself, have them draw the energy back to the starting point so that the experience feeds forward through the cycle, increasing in intensity. In the exercise we use the phrase "using imaginary hands."
The following language may be useful:

> Step right back into the place where you left off and feel the rush of feeling as you step back in.
>
> Imagine that you can reach out with imaginary hands and take hold of the best part of the feeling as it spreads through your body.
>
> Take hold of it and bring it back to the place where it started. Push it back through the center so that it doubles.
>
> Continue to push it out through your body and notice how it grows stronger. Grab the best part again and push it back through the center.
>
> Repeat this cycling, faster and faster until the state becomes surprisingly powerful."

- Remind participants to attend to the cycling of the feelings not the picture. It can also be useful to think of stirring or turning the felt sense.
- Remind the participants to focus more and more on the qualities of the felt state.

10. **Overload short term memory with impossible dimensions of feeling: location, texture, spread, depth, breadth, height, temperature, imagined color and imagined sound.** As the participants focus on more and more of these, the context and content will be crowded out of working memory and they will be left in a powerful, peaceful ecstasy that carries the flavor and physical tone of the original state. It is a generalized state of autonomic arousal that is framed by the original state.

    …And as you turn your attention …just gently turn your attention, … to the center of the feeling, you can begin to notice, … really notice… its temperature, … its color…. Notice whether it makes a sound … or a hum. And you can notice, really notice,… how the feeling moves…. Whether it is centered in your body, or beyond your body…. Whether it moves in a circle … or a loop … or a spiral…..whether it turns clockwise or counterclockwise … and whether it turns like a wheel …or like a turntable…. And as you notice the pattern of this movement, … you can reach out with imaginary hands … and begin to trace this movement… with those imaginary hands, … and if the movement of the feeling …is not a complete movement, … you can take those imaginary hands … and guide that feeling … through its own pattern, … back into its own center, … so that it grows …. and increases … and flows and multiplies. … And you can use those imaginary hands … to take hold of the feeling …. and move it faster … and faster … through its own center so that it doubles … and doubles again, … and grows stronger … and stronger, … and the pictures fade, … and the memories fade … and you find yourself floating … and resting, … down, … all … the … way … down, …into pleasant …. safe and ….warm. …Resting ….into your own ability … to feel …. good … now….

    Allow participants to remain in state for a while. They may safely be allowed to remain in this state for extended periods.

11. **Gently call the participants back to the present time and place.** This may be done casually, "Come on back. Reorient to the room and the present context in a way that is comfortable and that allows you to retain the lessons of this exercise in the present context. Come on back. NOW."

    An important part of the exercise is the abstraction of the feeling from the memory.
    - We begin with a remembered experience to gain access to a feeling state.
    - We enhance the memory to increase the felt sense of the experience.
    - We then focus more and more on the feeling in order to lose it from the memory and discover the feeling as something associated with the participant's own capacity to feel; independent of external influences.

After a brief discussion, repeat the sequence from 8 to 11 allowing them to discover **"How much you can enjoy that state."** And **"How many dimensions of wonder you can find inside."**

Script Two
Anchoring a resource state

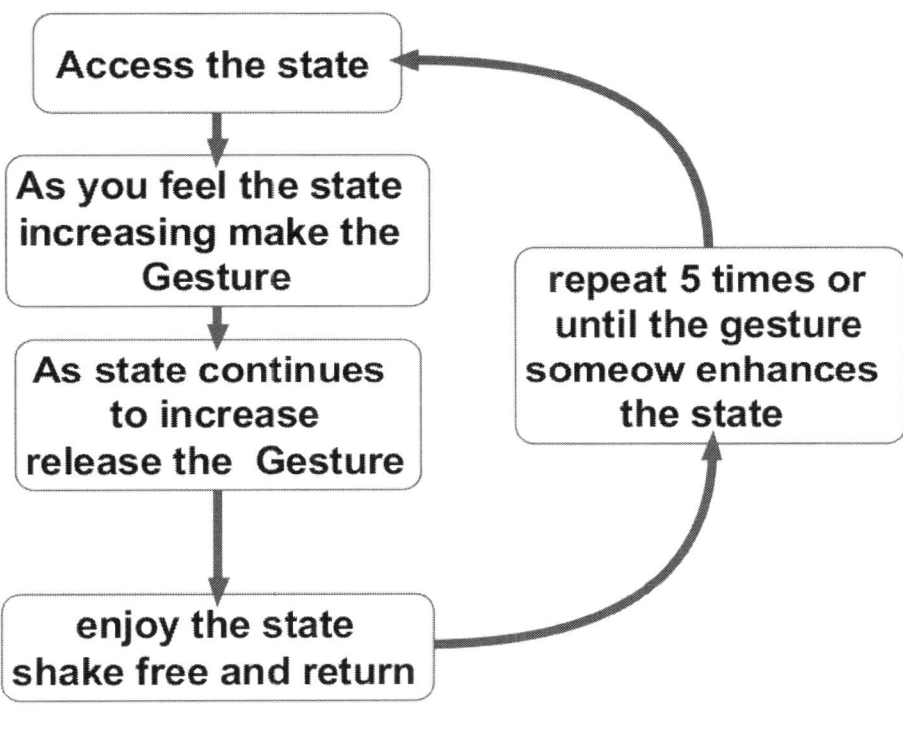

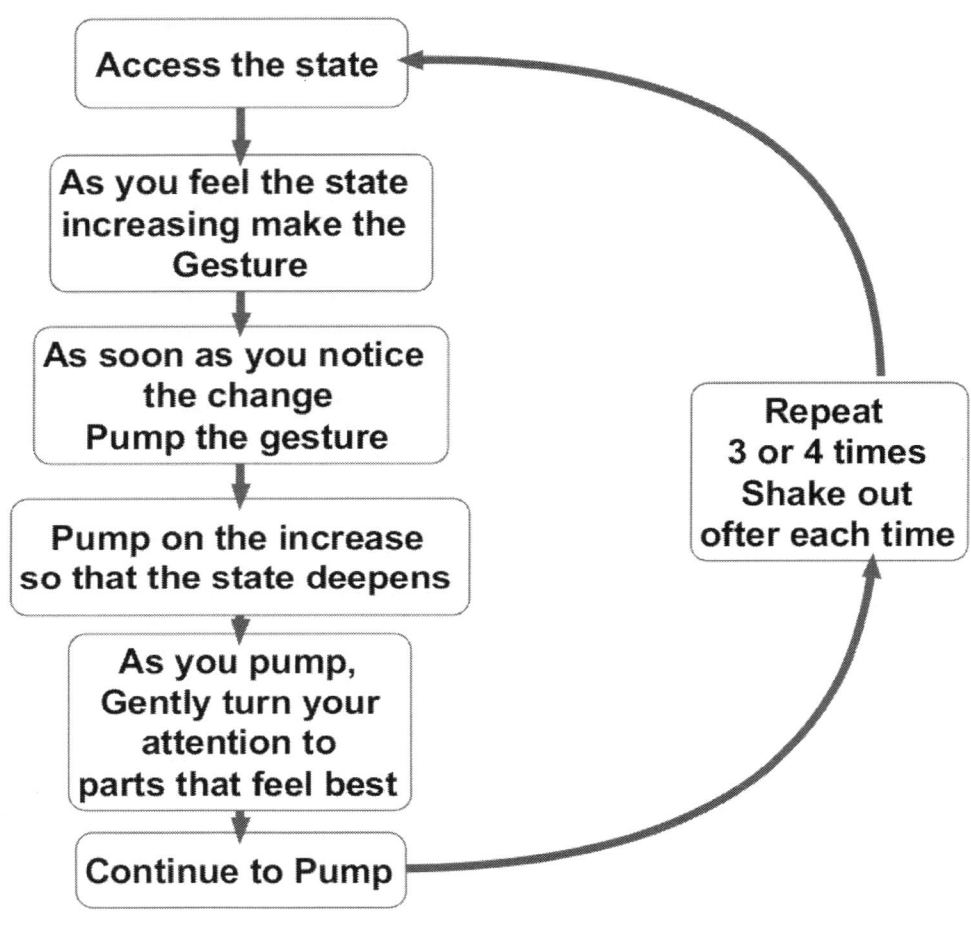

Scripts

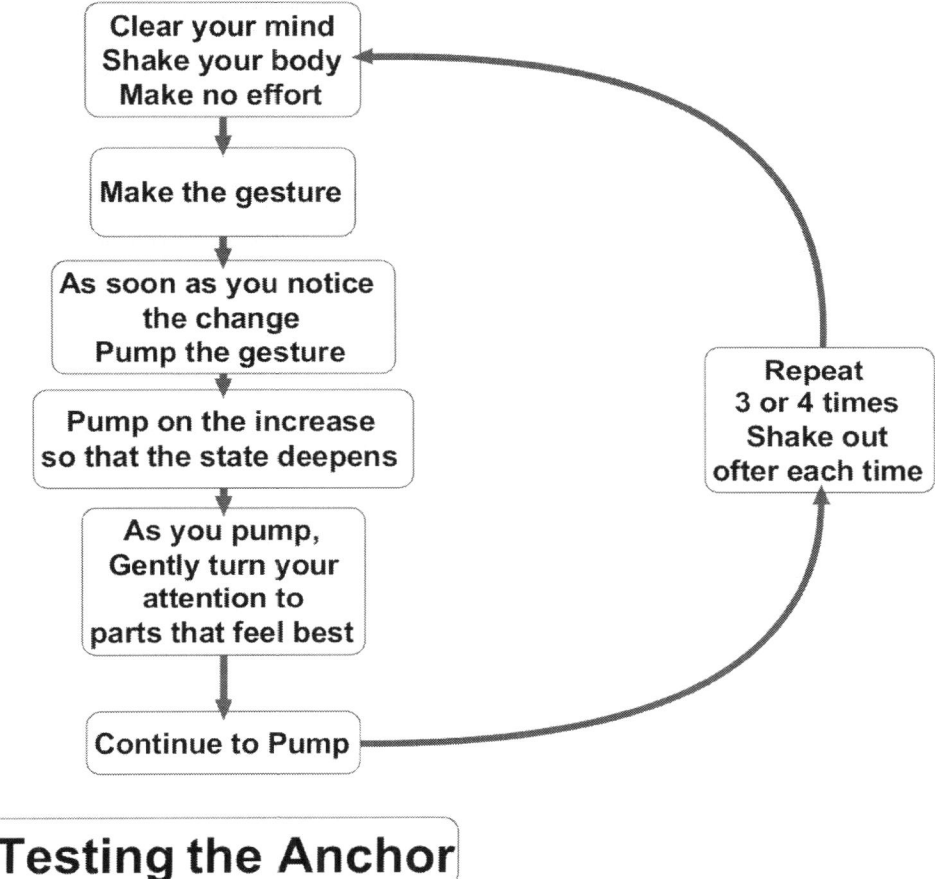

Testing the Anchor

Script Three
Creating NOW

Script for creating NOW

Begin by illustrating the gesture. A gently clenched fist made with the non-dominant hand. Practice a few times and remind the clients that they already know how to do this as they have already created and practiced other anchors.

Fire off Focus
and allow your attention
to rest down into the very best part.
Gently turn
your attention
into the patterns
of tension and relaxation,
the rhythms
of taking hold and letting go.
And pump it, and pump it and pump it.

Allow your self
to float down and back
into the very best part
and become aware of pattern
of the state,
the tensions and relaxations,
the height and depth,
where it is warm and where it is moist.
How it moves and where
it is very, very still…
Hold that in mind and body
as a single pattern of being
A pattern that you can remember
and rest down into it.

And as you hold that pattern
in mind and body,
move your fingers to fire off Solid.
And continuing
to hold the pattern
of focus in mind,
notice how Solid
arises in your body.
Ad pmp it and pum it and pump it.
Notice how it moves
and weaves
and interacts with Focus.
Notice how its patterns
of tension and relaxation,
taking hold and letting go,
Interweave
and merge
and combine with Focus
Weave them and harmonize them,
integrate them
and notice,
really notice
how they complement each other.
Get a sense of their dance
and the new pattern
that emerges as they
meld and join,
creating
something
new.

Notice how
they complement each other…
Notice their contrasts. …
Hold the combined pattern
in mind and body
as you move your fingers
to the GOOD gesture. …
Fire it off
and notice how the sensations of GOOD
dance with
and compliment
the feelings
already present. …
Become aware
of the changing patterns …
of tension and relaxation, …
warmth and light. …
Notice how they weave
together
into a new pattern.

Weave them and harmonize them,
integrate them and notice, really notice
how they complement each other.
Get a sense of their dance and the new
pattern that emerges as they meld and join,
creating something new.

Notice how they complement each other…

Notice their contrasts. ...
Hold the combined pattern in mind and
body as you move your fingers to the FUN
gesture. ...
Fire it off and notice how the sensations
associated with FUN
dance with and compliment the feelings
already present. ...
Become aware of the changing patterns ...
of tension and relaxation, ...
warmth and light. ...
Notice how they weave together into a new
pattern.

Fire off YES ...
and become aware
of its contribution ...
to the new patterns of feeling. ...
Notice how it rises in the midst, ...
creating a new space, ...
a new harmony, ...
a new pattern of feeling. ...
Become aware
of its temperatures and patterns, ...
its silences and sounds. ...
And, ...
as you notice,
really notice ...
how they all come together ...
in a new kind of way;
make a fist, ...
the anchor for NOW. ...
Pay attention to the new state
that is now arising. ...
gently pump or pulse the gesture. ...
Use the skill of pumping
 and anchoring
as you have been
practicing ...
Continue to become aware ...
of new facets ...
of the experience, ...
pumping as you go.
Explore the depth ...
and height ...
 and breadth ...
of the new state ...
and pump it. ...
Rest into it ...
and pump it. ...
Allow yourself to float. ...
And pump it, ... and pump it, ... and
pump it, ... and pump it. ...
Gently ...
turn your attention ...

to some wonderful part. ...
AnsPump it into the center.

Notice that you can
gently
turn your attention
to the very best part
and pump it
so it expands
to fill the center.
 Discover something new
about how much
deep joy
can be yours
as you pump.

Spend some time
enjoying and exploring
that state

Let the Gesture go
and come all of the way back.

Meditations

Introduction

Most sessions can end with a meditation. All of the meditations are scripted and, with minimal training, can be used by the group leader with a great deal of success.

In the context of our treatment Program the meditation provides several positive effects: 1) Self esteem. People like to think that they are smart. Teaching them the thinking strategies of the world's geniuses (as two of the meditations purport to do) is an ego booster. 2) Expansion of options. Addictions entail a narrowing of options. Any behavior or set of behaviors that expands possibility can be part of a strategy for combating substance abuse. All of the meditations enhance options. 3) Euphoric relaxation. Meditations can lead people into positive states of relaxation and non-drug induced euphoria. *It feels good.* 4) Convincer. As the experience of the meditations provides a profoundly relaxing and altogether pleasant experience, participants become more likely to attach positive feelings towards the Program and the facilitators. It is not uncommon for less cooperative participants to 'turn around' after one of the meditation experiences. 5) Peak experiences. Each meditation is chosen to help stimulate the continuing experience of Maslowian peak experiences. These are intended to teach the important lesson that **every** individual has personal worth and significance and that it is possible to attain extraordinary states of personal enjoyment without drugs.

Future Perfect capitalizes on the participants' access to positive resources states and uses them to provide an experience of a hoped-for future. At the same time it installs pleasurable associations with the use of the resources and skills learned in the Program.

New Worlds to Gain uses conflation of an imagined future experience of NOW with the present experience to provide a present-time enhancement of the felt-sense of NOW.

Please note that the meditations are written in script form. If you read them slowly, as phrased, pausing between the lines and attempting to pace them to the breathing of your students, you will get the best results. As you grow more adept at using them, you may wish to emphasize the words that appear in bold. Remember that every phrase should end on a *downward* inflection (Bandler, 1995, Bandler & Grinder, 1975b; Bandler & MacDonald, 1988; Bodenhammer & Hall, 1988; Dilts, 1993, 2001; Gray, 2001, 2002).

References

Andreas, C., & Andreas, S. (1989). *Heart of the mind.* Moab, UT: Real People Press.

Andreas, S. & Andreas, C. (1987). *Change your mind— and keep the change.* Moab, UT: Real People Press.

Bandler, R.. (1993). *Time for a change.* Capitola, CA: Meta Publications.

Bandler, R. & Grinder, J.. (1975b). *Patterns in the hypnotic techniques of Milton H. Erickson, MD, Volume 1.* Cupertino, CA: Meta Publications.

Bandler, R. & Grinder, J. (1979). *Frogs into princes.* Moab, UT: Real People Press.

Bandler, R. & MacDonald, W. (1987). *An insider's guide to submodalities.* Cupertino, CA: Meta Publications.

Bodenhammer, B. G. & Hall, L. M. (1998). *The user's manual for the brain: The complete manual for neuro-linguistic programming practitioner certification.* Institute of Neuro Semantics.

Dilts, R. (1993, 2001). *Strategies of genius.* Cupertino, CA: Meta Publications.

Gray, R. M. (2001). Addictions and the Self: A self-enhancement model for drug treatment in the criminal justice system. *The Journal of Social Work Practice in the Addictions.* 2(1).

Gray, R. M. (2002). The Brooklyn Program: Innovative approaches to substance abuse treatment. *Federal Probation Quarterly, 66(3).* December 2002.

Wenger, W. & Poe, R. (1997). *The Einstein factor.* Rocklin, CA: Prima Publishing.

Meditations

New Worlds to Gain

Presuppositions underlying the exercise:

This meditation is designed to provide specific experiences of the strengthening of the NOW response in an imagined future. By imagining NOW at a level of intensity it might have after practicing it for five or ten years, the participant creates a real time experience of NOW in the imagined future. The imagined future allows them to freely add dimensions that might be difficult without this mechanism for disarming the critical faculties. The meditation is wholly compatible with the state intensifications mentioned in Exercise Four.

Basic to the meditation are the assumptions that the participants have mastered NOW and that the facilitator has spent some time using the other meditations. It also assumes that the participants have developed significant expertise in developing and maintaining the altered states taught in previous exercises on their own

Instructional Notes

The meditation is much sparser than the others. It is a basic outline providing a few prompts for the participants as they enjoy the NOW state (or any other). Take your time and allow the temporal ambiguities to sink in. Ad lib using the verbal patterns presented.

Meditation New Worlds to Gain

PLEASE NOTE that the scripts are to be read column by column; each page should be read before proceeding to the next.

Fire off NOW.

As soon as you become aware of the beginnings of the feeling, pump it. Continue to pump it at the first hint of the rush.
Pump it up.
Make it very intense.

(Pause– wait for obvious indicia of relaxation and unconscious responding)

As the feeling intensifies.....
Allow yourself to float.....
Float into the feeling and...
As the feeling intensifies...
Find yourself floating....
Floating up and up.....
Up and up......
Out of your body.....
Floating.....
And pump it.....
And pump it,.....
And pump it......
(Pause)
Let the feeling .....
Take you .....
To a time .....
In the future .....
When .....
you have been using NOW .....
for five years or ten years. .....
You have been using it....
expertly. .....
Practicing, ...
Learning new ways to .....
Experience NOW.....
Finding new depths .....
in NOW......
Float into a time.....
NOW......
When NOW.....
has become.....
your normal.....
mode.....
of being.....
in the world......

See yourself.....
experiencing NOW.....
after using it for.....
ten years......
How do you look.....?
How do you breathe.....?
What is new.....?
And pump it......
And pump it......
And pump it......
(Pause)
Step all of the way into.....
The experience.....
Experiencing NOW, then.....
In the future......
Let the feeling.....
wash.....
over you......
Note something.....
about.....
how.....
it feels......
GOOD.....
What.....
new dimensions.....

Meditations New Worlds to Gain

are you.....

aware of.....

How.....

strong.....

How deep.....

Pump it.....

Explore it......

Rest into it......

And pump it......

Float into.....

New dimensions.....

Of NOW......

And pump it......

And pump it......

And pump it......

(Pause)

As you enjoy.....

This future.....

Experience.....

NOW......

Realize.....

Really realize....

You are.....

Experiencing.....

That future.....

NOW......

and that NOW.....

and then.....

and then.....

and NOW.....

Intermingle.....

and merge......

And because.....

You have.....

Experienced.....

That Now.....

NOW.....

It must be.....

Still stronger......

NOW.....

(Pause)

Focus into your future.....

NOW.....

And notice.....

some.....

new.....

depth.....

In.....

The future.....

NOW.....

And Notice.....

How Now.....

And then.....

And then.....

And NOW.....

Converge.....

And merge.....

Grow.....

And expand.....

together......

And pump it.....

and pump it.....

and pump it.....

and pump it......

And when you are ready......

When you know.....

Really Know.....

Certain things.....

That you do.....

Know already.....

Come fully back,

Refreshed and fully present.

NOW.

Meditations

Future Perfect

Presuppositions underlying the exercise:

We have already noted that many of the techniques used in the program focus on the use of Pseudo-Orientations in Time. This meditation is designed to enhance the present experience of future time. It assumes that the participants have mastered one or more states and that they have practiced some technique for enhancing the states.

The meditation is used to reinforce the experience of positive resources and to employ them as a draw into a believable future. It further provides an opportunity for the participant to experience positive states in contexts outside of the treatment situation and so seeds generalization of the feelings to other contexts.

Instructional Notes

Future Perfect begins with the elicitation of an Anchored state, preferably the NOW state. It makes use of standard enhancements to deepen the state and to move the participant into a desired future. The Facilitator should be mindful of the time it takes the group as a whole to reach state and ask for a signal, a small wave from each participant to indicate where they are before proceeding into the future. Remember that every change is keyed to the enhancement of the Anchored state.

Meditations
Future Perfect

PLEASE NOTE that the scripts are to be read column by column; each page should be read before proceeding to the next.

To begin with, relax, take a deep breath,
and fire off one of the resource states that we've been working with.

If you haven't anchored a resource state
or if you don't know which one to use,
think of an intensely pleasant memory,
one moment in that memory,
allow it to become very intense.
If it's a picture,
step all the way into it.

If it's an anchored resource state,
fire it off
and pump it up.
Let it become very intense.

When you've gotten there,
just enjoy it for a while,
Really enjoy it.
Continue to crank it up,
and give me a little wave so I know you're there.

(Pause while participants access the state.)

Now, as you pay close attention
to the qualities
of this pleasant resource state,
allow it to
continue to increase
within you.
You may wish to notice
the direction of it's flow
within.

You may wish to notice
the sequence of the onset,
the progression of the state.
Find the center of the state.
Find the very heart of it
and notice,
really notice
how it expands
and grows
and increases from there.
Find yourself wondering
how it would feel
if you allowed it to continue
to increase
for the next 5 days
the same way
it's increasing now.
Let it become twice as strong,
twice as intense.

Begin to notice
how it spreads and flows;
How it expands and grows.
Allow it to expand
through your entire being.
And as you
allow this feeling
to expand
through your entire being;
you can begin to feel yourself
floating up,
up out of your present posture,
up above the room.

As you allow this feeling,

this experience to continue,

to find yourself moving upward,

upward, floating

above the building.

Upward.

Allowing it to

increase and pulsate

to grow and flow

all the way through.

It lifts you up

and you can

see the city

or the town below.

And as you continue

to enjoy this state

ever more intensely,

ever more completely,

you could

perhaps

see your time line,

the line of your life

stretched out below you.

Down there is the past,

way back there

is the day of your birth,

and in the other direction

glowing and shining

full of hope,

see the tomorrows

that you have

yet to know.

And as you allow this feeling

to grow and increase,

to flow and to fill

every part of your being,

you can find yourself drawn

towards one of those futures.

A future in which

this feeling becomes

the normal aspect

of your life.

And as you pay attention

to the increasing of this feeling,

how it increases and flows

as it glows, and shines

as it sparkles ever more intensely

out of the midst of your being,

you can find yourself drawn

down towards that future

down towards that time line

to a place that shines

and glows with promise.

A place that attracts you

and reflects back to you

in an ever increasing manner,

the power of this energy.

And as you

allow this feeling

to increase and to spread

through you and your surroundings,

you find yourself settling

into that future.

And paying attention

to this feeling

it begins

to populate that future

with people,

with artifacts,

with all of the things

that let you know
that this is where
you live and
where you work.
That this is the time
and this is the place
where you have accomplished the desires
that you've been seeking.
And this feeling
is the natural feeling
of the life you've created
for yourself.
And allowing that feeling
to increase and resonate
with this place and time,
you begin to see
more and more details,
noticing
where you are sitting
or standing
or lying,
becoming aware
of the clothes you are wearing
of the people who surround you
of the people who support you
who have supported you
all along.
And as you look
at the things around you
the feeling intensifies
and as they gain clarity
you begin to know,
really know
who you really are
and what you do
in this future,
and once again,
who is with you?
And with a certain amount
of familiarity and clarity
of comfort and thankfulness,
you begin to realize
just where you are
and when you are,
and what you've been through
and what you've accomplished.
And as you take the time
the time to remember,
really remember
the steps that you took
that brought you here,
perhaps you can remember
the last thing you did.
The thing that made
all of this possible.
And with a certain sense of familiarity,
of comfortable confidence,
relaxing in a way
that's right
for you,
you can begin to remember
all of the choices
that you made,
the decisions,
the actions
that you took
that made all of this possible.
And as these feelings increase
in familiarity and comfort,
as you begin to recognize
how much effort
you've put into
this well-deserved future,
you can, perhaps

remember the steps
that you took concretely,
one at a time,
beginning back there
in the early months
and the first few years
of the 21st century.
And there's a certain pride,
a certain sense of accomplishment
that adds to this
increasingly powerful sense
of who you really are.
As you being to really remember
the choices and the struggles
the actions and the victories
that brought you to this moment.
And that feeling
that glows and pulsates,
that shines and sparkles,
that increases,
that fills this future with a sense
of incredible pleasure,
can begin
to lift you up
and connect then with now,
and now with then,
bridging between future and present,
the past and the nows
that all come together
in this possible future
perfect place.
And again
floating upward
with ever increasing
senses of pleasure and confidence,
of renewed strength
of this feeling.
You begin to float back
into the present time,
carrying with you
the realization
of all that you are capable of
in such a short time.
And as you find yourself
floating back
into your own body
and into the present time,
you can feel a certain certainty
knowing who you are
and who you are becoming,
and allowing that feeling
to expand and increase
through your entire being
until you feel
much, much better
than you have in a long, long time
knowing the futures that await you
and ready to begin making the choices
that are yours.
You can return to the present
feeling wonderfully well
and ready to begin
NOW.

Application Notes

These application notes were originally prepared to assist an implementation of the Brooklyn Program in a South African prison context that never came to fruition. As time went on, I began to modify the notes for the use of others who wanted to facilitate the program and needed some further guidance.

While the manual generally moves from exercise to exercise, each of the notes is aimed at a session, several sessions or specific issues. They address many of the practical problems connected with running the program.

The best preparation overall, is to try out the exercises on yourself or perhaps to work them through with a few friends. If you know how they feel for *you*, you will be a much better guide. Most of my training is designed to do just that, provide the experience to the facilitator so that they can speak from personal experience. While everyone's experience of the exercises is unique, there are broad commonalities.

I have used the program as a substance use treatment program, but it is not limited to this use alone. The original application notes envisioned its use as a prerelease program in South Africa. With a prerelease population, or even a general inmate population you will probably want to spend quite a bit of time on Exercise Ten working it through multiple scenarios and learning to use the Smart Outcome Generator with and without the imagined futures that flow from the remainder of the program. I think that it would also be very useful as a spiritual orientation program for reasonably stable persons entering an institution or program of any kind. It might also be used with regular, non-criminal populations.

Wherever I have had the opportunity to train professionals in the use of the program, it has had the effect of awakening people to personal direction and a positive sense of self. It also has a way of precipitating changes in conscious life-priorities. It is not uncommon for people to discover that there are issues that they need to deal with and the program often awakens the motivation needed to tackle those issues. This often sets in motion life transitions that are not easy and which yield no simple answers.

Application Note 1

In every context, from the outset, emphasize that all of the techniques are easy, all of it is natural, it happens to all of us every day. We are only learning how to do it on a conscious level. The most important words that I use is "Notice ...just notice." This implies an easy observation that takes little or no effort. It is a simple turning of the attention towards a feeling or sensory percept. Avoid the word try. The word 'Try" implies failure. Simply instruct.

As the participants start on the first exercise remind them that the process is gentle. Suggest from the start, that if attention wavers, it can be gently returned to the memory. Things will be much easier as the wanderings are incorporated into the process as reminders to return to some facet of the memory. Like mindfulness meditation, the goal is to calmly return to the object of attention. Remind them that wanderings are normal and that for most people, except those well practiced in meditation, there is no such thing as a purely monolithic attention. As they practice the exercises and the states become more intense, they will have no problem with maintaining attention.

One of the advantages of focusing on a subjective state and its submodality structure is that the stimulus object has multiple levels. Even if the attention shifts from visual to auditory or kinesthetic elements of the object, there is always another submodality distinction to be discerned. This is helpful in training attention.

Exercise One is somewhat cumbersome as written. It is designed to provide some background for the program in general. Much of it, beyond the summary at the end, is unnecessary. In practice I present a very small number of submodalities in the exercises themselves. It runs largely as presented in the summary. Let them play with the complexities of it as part of the homework. There, they can find the combination of submodalities that works best for them. Encourage this. I don't usually hand out the written materials until after we have gone through several examples. After you have led them through several examples, the written exercise will make much more sense. Moreover, it will not serve as a distraction while leading the exercise.

As I present the exercise, I will sometimes poll the group after each change (What happened? What was different in your experience?), at other times I will go through the whole sequence (as in the summary) and then ask for responses.

As you lead the exercise, it can be very useful to have them observe how one sensory change affects the intensity of the other senses. I weave the modalities back and forth to make the underlying process part of conscious experience. Remember to give them time to respond between each suggestion. Be careful to notice how the majority of the population uses their sensory information. With a few questions you can ascertain whether the group should start with visual, kinesthetic or auditory information. If there are people raised in cultures that are radically different from ours, the, they may be particularly auditory. But, you never know until you ask.

Following the typical (American) sequence of visual, auditory and kinesthetic ... After going through the basic visual elements (bigger, brighter, closer), you might ask them to turn on the sound and turn up the volume "...and, as you do, you may notice that the picture grows sharper.... Notice the directions from which the sounds come.... And whether they are voices, music or just sounds... And as you notice that, you can become aware of how much more present the feelings become. Notice the interaction between the sounds and the pictures and begin to notice how it feels to be … really be... in the memory." As you loop them through similar language you will bring them to the place where the emotional feelings that characterized the original experience become quite strong. Have them notice how the intensity of feeling enhances the other elements of the memory. "And as you find yourself enjoying these feelings, notice how much sharper the whole experience becomes. Begin to notice new details of the sounds and sights. And as you notice that, just notice that the feeling itself grows stronger. And as the feeling grows..."

When it becomes obvious that people are moving more deeply into the memory (they will relax more, you will see changes in muscular tonus, increased facial symmetry, facial flushing, swelling lips and, movements that suggest the reliving of an experience), begin to turn their attention more and more to the details of feeling: temperature, direction of spin, orientation of movement, internal brightness, colors and vibration. Some of these are purely metaphorical. Have them connect with the texture of the feeling and how it spreads. Ask about its thickness (viscosity). As this goes on, remind them that the feelings will first enhance the memory, then, they will notice that as the feeling increases in intensity, the memory will begin to fade and they will become more and more aware of floating, just floating in a pure, positive, ecstatic state. Encourage them to rest into this, let the memory go and just enjoy the state.

At this point there will be objections to losing the memory. Let them know that the memory is only a vehicle. As we proceed, the feeling that they are learning—without the image and context—will be much stronger.

As you pass into the part where the felt experience is rotated, there are a few interesting and perhaps useful metaphors that come to mind. You may be familiar with the idioms "stirring things up", and being "stirred" by emotion. At other times people talk about feelings and responses that sent them "spinning," or "reeling"

I've been thinking about spinning and stirring and churning and grinding with mortars and pestles as a metaphor for the same kind of spinning that Bandler takes from Yoga. I'm suspecting that much of the mystical practice associated with alchemy and Wicca in general may have depended on just such an internalization of movement. Imagine the action of stirring a cauldron with a large spoon that requires two hands. Move the shaft inside yourself and begin to stir with two imaginary hands. Sometimes the hands move together and their motion describes a cylinder. Sometimes they work opposite one another, pulling with one and pushing with the other. In this case their motion describes an hourglass. Notice the difference that the locality of the top and bottom of the shaft makes. Change the length of the spoon to create different patterns of rotation in different Chakra systems. Move your hands so that they define different centers of rotation and different sizes of internal arcs. Upper and lower arcs may be of equal or different sizes. Notice how it changes when you imagine the left hand over the right and the right hand over the left. The application of this to the exercise should be apparent.

Be sensitive to your audience with these metaphors, some people may not find an association with witchcraft or non-western faiths useful. I usually keep far away from specific religious or spiritual content until they begin to have experiences that are of a significant enough level that they spontaneously make the identification.

Please remember that none of this is absolutely formulaic. Find a set of patterns that works for each group and play with it. If one way doesn't work, try another. All in all, the most important variable is depth of experience. The more powerful the altered state, the more convincing the exercise: the more pervasive the change. Milton Erickson used to leave people in trance for long periods to work out their issues. I like to see slackened jaws and long response latencies and slow returns to normal consciousness. Challenge each group to find out how much they can enjoy the states.

Application Note 2

I'm not sure that this is necessary for you, but linguistic patterns need to be considered as you do the Program. One of the emphases is, of course, the positive emphasis. I also try to keep it fast paced and light. There is no preaching or scolding, no sermonizing about the evils of drugs or crime. Persons in the system have had more than their share of that and, for the most part, it still doesn't work.

As for the exercises themselves, although the states are closely related to the hypnotic state, I have chosen to emphasize the distinction between what we do and hypnosis. This avoids all of the superstitious nonsense that still adheres to hypnosis. While people are learning how to access the states, I use a normal voice with an instructive, command inflection (see below). I try to make sure that the early instructions are clear, that participants have enough time to respond and that there is no hint of anything going on, except instructions and responses. Once people are moving into state, I will then begin to alter my tone, quiet my voice and use hypnotic tonal patterns to deepen the state. An important reason to avoid the idea of hypnosis is that part of our aim is to create a sense of personal control. The superstitions surrounding hypnosis often muddy this sense of personal control. As the participants build up the ability to access the states, the autonomous nature of the states becomes more self-evident. I keep repeating that our most important goal is to "give you back to yourself."

When the question of hypnosis comes up, I state categorically that this is not hypnosis. It has some similarities but it is not hypnosis. I describe it as much like meditation, or as one of the possible set of altered states of consciousness along the aesthetic spiritual continuum described by Newberg and D'Aquili.

There are quite a few linguistic patterns that can be useful. Don't strain to use all of them all of the time. If you present good information and keep it interesting, that always helps. Try a few of the patterns each week. Find the ones that work best for your audience. Remember that the issue is to get them to go as deep as possible into state. Sometimes just getting them going and leaving them to their own devices works best.

For deepening purposes, cognitive overload is very useful. Remember to suggest that they turn more and more attention to the kinesthetic aspects of the experience. Add more and more qualities of feeling, like brightness, sound, vibration, orientation, temperature, texture, etc. until all of the other senses run out of room in short-term memory.

From time to time people raise the question as to whether you can get stuck in these states. The answer is no. The worst that could happen is they would drift off to sleep and entertain the rest of the group with their snoring. One hypnosis text describes hypnotic coma as a state so deep that subjects become unwilling to respond. The remedy is to suggest that if they don't come out, they will never be allowed to reach that place again.

Others have asked if it can be habit forming or addictive. I say yes, I certainly hope so. This is what your brain is designed to become addicted to: Positive behaviors that increase flexibility and personal control. For those who are concerned about abusing the states, remind them that those who overdo it usually end up in monasteries.

From the very start, it is important to use the command tonality when giving instructions in the exercises. This means, that every instruction should end on a downward inflection. In general, questions end with an upward tone, boring facts with an even tone and commands with a downward tone. Start by making sure that every instruction ends with a downward tone.

A monotone is bad form. I always start with a normal level of emotional expression in my voice and increase it as the states deepen. After a while, as people respond, your speech can become very slow and you can use the kinds of emotional tonalities you might use with a young child or an animal. This is based on the idea that the unconscious responds to emotional cues.

You can also mark out instructions by altering the volume of your voice "You can NOTICE, REALLY NOTICE, Bob, how good it feels to FEEL GOOD, NOW..."

As you seek to deepen the states, you can use language like "and you can float... all... the way... down." while your voice decreases in pitch, so as to imitate the downward motion. You can even, in these places, begin with your voice aimed at the ceiling and lower it until you are speaking to the floor at the word down.

As noted previously, I avoid the word 'try'. While guiding them into state, all of your statements should be positive—about what they must perceive, or the direction in which you want their attention to turn. As you GENTLY TURN your attention towards ..." I emphasize words and phrases like "notice" and "Gently turn your attention." Once they are in state, there is little that needs to be done beyond continuing to offer deepening suggestions. The depth of the experience is a crucial piece. The more you encourage them to "Discover just how much pleasure you can find..." And "How many dimensions can you discover..." Or "You may notice mists, or landscapes, it may be dark or light, some people describe rainbows, others, the impression of space." the better. "NOTICE, that as you just LET GO, you gain more control over the depth of the state."

Any good book on Ericksonian hypnosis or NLP will have these linguistic patterns in them. You may want to take a look at *Frogs into Princes*, or *Trance Formations*, both by Bandler and Grinder. On the other hand, I've had a facilitator reading this right off the paper with no NLP skills and he's done quite well.

Another important part is pace. At the outset, I start quickly. As they get more deeply into state, my pace slows significantly. It becomes slow enough that every word or phrase becomes ambiguous and drives the states deeper. As you are doing it, it will sometimes seem painfully slow. As long as you are rewarded with slackened jaws, relaxed facial muscles and loss of conscious-style movement, you're doing well.

Application Note 3

As follow-up to the linguistic materials, unless you have already done some hypnosis, you might want to start by reading the Tesla Meditation (The first meditation) to some friends [this is no longer included in the materials. RG]. It will provide some nice practice in timing, ambiguity and tonal shifts that are fairly well scripted as written. I no longer use the meditations, but they are a good way to get used to the language patterns and cadence of trance.

Exercises One and Two merge together very well. I usually spend most of the second session going over several positive memories using the first exercise. I will often go back to the ones we used in the first session to jump-start the exercise. If you always start with something well practiced, your results will be better. Remember, the sooner that you get the physiology moving towards the positive altered states that we are inducing, the easier things will go.

Again, for the first exercise, you may or may not want to go all of the way through the submodality lists. I abbreviate them rather quickly.

There are a few objections that arise with some frequency. The obvious complaint is that this is nothing different than daydreaming. On some level this is true, but there are important differences:

1) This is *directed* day dreaming. It is teaching the skill of focused attention by consistently returning to the specific object of attention—the best few seconds of a remembered event. It is designed to make the experience dependable, accessible and controllable.

2) It is *structured* daydreaming. Rather than just going off in a reverie and following it wherever it goes, these techniques allow the participants to systematically recreate the felt sense of the remembered resource.

3) Using the brain's own methods for recreating the feeling event, we can control the quality of the experience on a conscious level. These same skills can be used consciously to enhance the quality of any memory in a systematic way.

4) One of our main goals in the exercises is the creation of the felt understanding that emotions are skills. They are not just events that are imposed upon us from without. They

are separable from the stimuli and events that evoke them and can be created at will. Unlike daydreaming, we use the memory to evoke the feeling and then enhance the feeling to the point where the memory disappears from consciousness and we are left floating in the intensified feeling.

5) Unlike daydreaming, these are practiced skills that are foundational to the creation of other experiential skills in later exercises.

Another objection is the felt hypocrisy of trying to force one's mind to remember a pleasant memory in the presence of a negative stimulus. When this is raised, it is a perfect time to underscore the importance of learning to do the feeling independent of the memory. Agree with the complainer and point out that 1) we use the memory to get to the feeling, but once we have the feeling, it is important to let the memory fade. 2) Acknowledge at this time that it is very tempting to hold on to the memory data because it is our primary means of orienting to the feeling. Assure them that as they let go of the pictures, sound and context, the felt sense that follows will be much deeper and much more useful. Once the feeling is separated from the context and becomes a skill that can be called upon at will, they will understand how important it was to lose the content.

As you guide the participants through several examples from exercise one, be sure that you emphasize that they allow the memory to fade so that only feeling is left. Your goal is to teach as many of the participants as you can, as quickly as you can, how to get to a deep, ecstatic, meditative state. When people start to be surprised at the depth of the state and have some difficulty coming back, you have begun to get them where you want them.

When the participants have really *gotten* it you will notice most of the following:

1) They will be able to describe—or point to—a sequence of internal responses (e.g., warmings, coolings, churnings, turnings, sensations of weight or lightness, etc.).

2) Physiological signs: Changes in posture, facial expression, heart rate, breathing and skin tone. Many will begin to express rhythmic movements that reflect the underlying experience. Although they differ from person to person, state changes will be observable. They will relax more; faces will become more symmetrical; movements will be minimized.

3) Response latency: Persons who have entered the deep states required will either not respond to external stimuli (loud noises) or will respond with marked latencies (eye movements several seconds after the sound).

4) Perseveration: Persons who access the states appropriately usually take some time to return to normal consciousness. Persons who immediately return to normal voice tone and reaction time are suspect.

5) Mood change: People who begin the session in negative states quickly change to more positive affects. Persons who retain a negative mood are suspect.

Once people have gotten the idea of how to get into the states, introduce the second exercise. You can tell them that these states will become tools that they will use to turn on and off a series of specific states as they will.

I usually just go through the definitions and poll the group for some examples before they go. It is important that they choose appropriate exemplars. Although ultimately mooted by the depth of state, I still like to insist that the exemplars, be clean, sober and legal examples.

I sometimes wonder about the ultimate utility of the names that I've chosen for the states. It makes it convenient to name each state, but this also means that you have to keep pointing people away from the name tag and back to the definition. Remind them to go by the explanatory definitions and remind them that these should all be pleasant, positive memories.

Again, the most important criterion for all of them is that they should be free of excess emotional baggage. These are best chosen from past events that are acknowledged to be over or complete in themselves. Maudlin reminiscences should be avoided.

Focus. Aside from the criteria already presented, make sure that it is a pleasant and empowering event. Young infatuations are fine, as is sex, as long as you can get them to give up the pictures and other contextual data. People who have had experiences with meditation may use this here. Focus tends to attract current time concerns. Do not allow a current issue, or the present time need to get focused define the state. Find something finished, pleasant and powerful. Think of Flow states.

Solid. Again, do your best to move everyone into a place where they made a choice from a field of multiple options. I like the example of buying a suit or a pair of shoes. There is a series of finer and finer decisions made in succession: the store, gender, style, cost, color, fit, feel, etc. By the time the transaction is complete, there have been a whole series of decisions made. The process ends with a small felt sense of "This is it". Emphasize that this is often a small feeling, but it is the one we want. This is a good place to emphasize the idea that one of the skills we are learning is an emotional magnifier. These tools can enhance a relatively weak experience and produce a powerful state of being.

Good. This state is often confused with satisfaction or success. What I'm after is the sense of "getting it" when a complex set of individual acts coalesces into a single coherent movement. I often use the feel of making a stick shift work the first time, or the first smooth ride on a bicycle. A better example is when a set of musical notes and the movements that create them shift from being a set of awkward, individual movements into a smooth, single motion. Dancers and athletes have the same experience. I like to suggest the moment when a speaker of two languages moves from word by word translation to smooth comprehension.

Fun. just that, fun. Any pleasant experience will do. Again, there should be no regrets, no pining for the old life; just an appreciation of an experience of enjoyment. A lot of our offenders seem to stumble on this one. This is another place where you can emphasize that we can start with something very small and magnify it.

Yes. Yes is a moment of confident appreciation of some well-practiced skill. Whether the appreciation comes in present time or was a past realization of the quality of work or skill is relatively unimportant. We are looking for a well-grounded sense of accomplishment and confidence. People often find this one leaving them a little cocky. That's just right.

I usually review these definitions once through and then poll the group, one category at a time, to ensure that everyone gets it right. People often want to insert their own choices for the states independent of the criteria. Insist on these criteria. Also insist that the exemplars chosen have nothing to do with drugs, alcohol or prison. For our purposes the decision 'to stop using drugs' is not a valid decision. It lacks the process that we seek and it is ill-formed according to the standard NLP well-formed-ness conditions. Likewise, for our purposes, the day someone got out of prison is not an example of Good.

Once the exemplars have been chosen, work through one example with everyone. I often use Focus, Solid or Good as the first one. Use the exact techniques practiced with Exercise One. The only difference in this exercise is the categories of experience are predefined and the homework requires a little writing to stabilize one memory for each category. Challenge them to discover how quickly they can lose the picture and how many kinds of enjoyment they can discover in each state.

When they return the following week, spend some time asking how they did and let them describe their experiences. Handle any problems, and explain any difficulties. Spend most of the rest of the class reviewing the states. Make sure that everyone has examples for as many of the five as possible. Have them step into each one and spin up the felt sense of the experience to ecstatic levels. Make sure that you have enhanced as many of the exemplars as possible.

Before they go, show them how to create an anchor. Have them all access the same state. Let them access it either by going through the process developed in exercise one, or by going right to the feeling and beginning to spin it. Give them time to spin it to ecstatic levels. As they do it, gently remind them to turn more and more of their attention to the feeling: "Notice how deep it goes..... What direction it moves in... is it clockwise or counter-clockwise.... Does it move up or down, Left or right.... is vertical or horizontal...? How thick is it... Does it have temperature.... notice the textures of this feeling? And as you really enjoy this... feeling more and more... just let the memories fade... and the room can fade... and I can fade....."

When everyone (or nearly everyone) is in deep trance, gently invite them back. Have them shake out the state and tell them to listen carefully. Show them the appropriate anchor and tell them to listen before doing anything. Tell them as follows:

> In just a minute I'm going to ask you to close your eyes and access the state. At that point, I want you to do it very quickly and spin it up to a wonderful level. Do this as fast as you can. When you are really aware of the feeling, make the gesture. Hold it for just two seconds. Let the gesture go while you continue to enhance the state. Enjoy the state for a few minutes, then shake

the state out, open your eyes, and look at me. Show them the gesture again and let them do it. You may or may not want to talk them through it.

When everyone has opened their eyes, or enough of them have so that you are assured that they have had enough time, call them back and give them the same instruction. Assure them that right now, the gesture is meaningless. What they are learning is how to attach the feeling to the button, or gesture, or anchor. You may even suggest that for the first several times the gesture will be a distraction. Repeat the instructions and have them do it again. When you call them back, instruct them to do the same thing two or three more times or until something changes when they make the gesture.

For most people, five repetitions are usually enough to create the conditioned link between the gesture and the felt state. Some will take more, some will take fewer. What should be happening now is that as you poll the group, people will talk about changes in perceptions. For some the feelings will simply intensify; for others, the internal movements quicken. For those who retain the content of the memory, the image speeds up or the picture clarifies.

At this point have them repeat the exercise again, but with this difference: as soon as they experience some difference as a result of making the gesture, make it again. And as soon as they note their body/mind responding to that do it again. Have them pump the gesture in this manner until they take it as far as they can. Then have them take it farther.

As they pump the gesture there are a few things to remember:

1) It is not necessary to take the fingers apart. Pumping seems to work best as a tightening and loosening of the hand rather than a complete opening and closing of the gesture.

2) It is important that they make the gesture at the first instant of noticing a response or change. "The minute that you begin to notice some change in your body related to the gesture, make the gesture again. As soon as you notice a change from that, do it again. Don't wait for it to grow strong. Pump at the slightest inkling of change."

3) Some people will report that the experience flashes or strobes. If you watch them, the strobing matches their exaggerated opening and closing of the gesture. Suggest that they rub or pulse the gesture and that they speed it up so that the feelings grow smoothly.

4) Suggest further that they notice the connection between their fingers and the feelings. Remind them to notice the rhythm of the connection and that they will find a time when, just by speeding up the gesture, the feeling will grow.

I usually hand out Exercise Three at the end of the session and instruct them to anchor as many of the states as they can. Before they leave, I show them the gestures and point to the place in Exercise Three where they are described (at the end of each descriptive paragraph). I also remind them to use my gestures, as these will be one of the criteria that I will be testing them on during the one-on-ones.

Application Note 4

Now we are officially into the anchoring piece. By now, almost everyone should have identified a positive resource for each of the categories and has had some practice at least stabilizing the memory. From here on in, most of the next few weeks will focus on practicing the states and the technique of anchoring.

The first thing that you do in this session is just poll the group to find out how everyone did with the homework. Some will have anchored all five; many will be no farther than where you left off last week.

After getting a sense of where everyone is and answering questions, invite the group to access focus. Tell them not to use the fingers yet, but to go through the process of accessing the feeling and cranking it up to ecstatic levels. If everyone cannot yet get to the felt sense directly (ie, without the basic steps from exercises one and two), invite them to go through the process of enhancement until they can spin the state to ecstatic levels. For those who can access the feeling directly, have them spin it until they float. Give the group enough time to get there. With some groups, before proceeding, I will do a quick review of all five states, allowing three to five minutes for each.

Follow this procedure again, without the gestures for one of the states, and have them notice how fast they can do it now. Call them back and almost immediately send them back to the state, pointing out how quickly they can experience the fullness of the experience and suggesting that they notice the "rush" as they return to state. Call them back and send them in again with the suggestion to spin the rush into the feeling and enjoy it. Do this once more and give them a very short time (a minute or so is often enough) before calling them back. Show them the gesture for the state and tell them to do the same thing, but now, have them make the gesture as soon as they BEGIN to feel the rush. Have them all repeat this step at their own speed. Have them do it five times or until they begin to feel a difference when they make the gesture.

Each time they finish a segment, let them stop and look at you for a second. When they have gotten to the point where they perceive a change as a result of making the anchor, then they should do it once more and then stop and wave at you, or otherwise let you know they're done.

After the group is largely back—you really can't wait for everyone—have them do exactly the same thing—step right into the state, as you feel the rush, make the gesture. For this time, have them notice the moment when the experience starts to change, just as they make the gesture. As soon as they notice the beginning of this new rush or change, have them remake the gesture (pulse it), as soon as they begin to feel the rush from that, do it again, and again, and again. Have them continue pumping/pulsing for three to five minutes, or until they are all immobilized. Then call them back, discuss their results and do it again.

I've noticed that when this is working well, there is a profound stillness in the room. Except for the sound of breathing, there will very little motion and very little sound (there is the occasional snorer). People, who have not gotten it, will clearly stand out against this background.

After the second round of pumping, bring them back and poll the group for responses. Deal with people for whom the state flashes (not pumping fast enough), for whom the state diminishes after getting strong (stop waiting for the state to get strong, pump immediately on the first inkling of change) and those for whom it still doesn't work (review the process at home or see me in the office). After they have discussed their experiences, spend a minute pointing out, the automaticity of what they've created. Talk a little bit about how this is like a button. You press the button and get a wave of feeling. Press it fast enough and you get a smooth tone.

Now it is time to test the anchor. Tell the group to do nothing until you tell them and instruct them to sit straight in their chairs with their feet flat on the floor (This is good practice from the beginning. It also keeps sleepers more awake.). Without making any other effort, have them close their eyes, make the gesture and begin to pump it as soon as they feel anything in their bodies. Remind them to catch the first hint of change and to pump so fast that there is no opportunity for anything but the feeling—no matter how small— that they have noticed in their body. Let them continue pumping until they are all immobile (except for the gesture) and deeply relaxed. Bring them back and debrief.

Now go through this process of anchoring and testing, for all of the states (or as many of the states as you can get to). Assign the remainder to be done as homework.

As to the sequence of the anchors... Before they go, show them how the anchors run, in the same order as in the written exercise, down the first three fingers of either hand: Tip of thumb to tip of index finger, tip of thumb to first joint of index finger, tip of thumb to tip of middle finger, tip of thumb to first joint of middle finger, tip of thumb to tip of ring finger.

As you work through the exercise, you might, both for this session and next, use an order different from that suggested in the exercise. In the second week of anchoring, I will often start with yes, then do fun and good. It is not uncommon to spend two or three weeks getting the anchors down.

As the participants get more into the altered states, after they've done one or two, in this session and the next, challenge them to find out how fast they can do it. Typically, the faster they get it, the better it works. You can easily spend between two and three sessions working on the anchors.

Recently, I have stopped using Exercise Four, Keys to Enhancing Subjective Experience, as a discrete session. What I do is integrate the best three or four techniques into the patter that I use while they practice the anchors.

For example, once they are really getting into a state, I will suggest that they "Really notice, just notice and enjoy, the speed with which THIS FEELING IS INCREASING, and AS YOU DO, you may wonder, REALLY WONDER, how it would feel, if it kept INCREASING LIKE THIS for the next five days.... in the next ten seconds. Let that happen, now." This is number four from the exercise

Number eight is quite useful and is based on the standard NLP swish pattern. Again, at a time when they have all gotten into that still space, suggest that they "Imagine, REALLY IMAGINE, that there is a small spot opening in center of your field of vision. As you CONTINUE TO ENJOY THIS state of mind and body, imagine that this spot is a picture of you, ENJOYING THIS STATE 50 times more intensely. As the picture becomes clearer, NOTICE.... how you look.... and FEEL in that image. NOTICE the LOOK ON YOUR FACE... And STEP ALL THE WAY.... INTO THE PICTURE... and let this feeling wash over you. Do it quickly....." Wait a while and do it again.

I'm no longer using the dials and knobs but you can try them out. Some people find them very useful, some don't.

Number ten is a standard. When first reviewing any anchor, I will usually use this one. I will also make sure that the anchor is tested afterwards and remind them that this is just a way to enhance the state when they are starting to practice.

By the fourth or fifth session, I'm usually pretty sure about who is doing well and who isn't. I now begin to invite the non-participants and those who are having trouble into my office for a one-on-one to determine whether they need help, or if they need to start over.

When all of the states have been anchored, I typically assign a version of Exercise Six, Pacing the Future. Each participant is instructed to practice one anchor each day. They are to start off each morning with the anchor and end each night with it. Upon arising, after firing off the anchor and from within the anchored state, they are to identify three places during the day where they anticipate the feeling of the anchor will be useful. They are to write these down and give them no further attention. The intent is to produce an anticipatory set for using the anchored state in those contexts. They are also required, before continuing the day, to make three appointments with themselves through the day. At each of the appointed times (they should vary from day to day) they will: 1. Fire off and enjoy the anchor (eyes open or closed, privately or publicly). 2. Take a few minutes to review how well the day has gone to this point. At this time they will make a few notes on their accomplishments and then get on with business.

At the end of the day they will again fire off the anchor, review the day and make some progress notes. Finally they are instructed to drift off to sleep firing off the day's anchor.

The intent here is to create connections, to seed generalization of the positive resources throughout their days. This overcomes the problem of having a skill that is limited to the practice hall but is otherwise unavailable. It also provides an opportunity for them to reflect on the value of the techniques in real life.

I always warn them that this exercise will be collected. Sometimes, when I cannot be present, I will extend the exercise for two weeks and have them send in the first weeks' worth. Most people enjoy the exercise and produce good stories from it.

Application Note 5

The One-on-One session

After people have had a few sessions practicing the anchors, usually about the time when all five anchors should be fully functional, I begin to bring in the least advanced participants for one-on-one sessions. This has several purposes. 1) It sends a subtle message to the other participants in the group that this is serious business and that they'd do well to practice the skills. 2) It allows me to find people who need extra help and get them up to speed. 3) It allows me to find the people who are just "doing time" and eliminate them. 4) It prevents the slower students from holding back progress in the rest of the class.

During the normal process of leading the groups you will have been watching the responses of the participants. By now, you will have noticed that there are people who are just not 'getting it', and some who are not even trying. These are the people that we want to target first. They are already identifiable by various behaviors (Don't memorize the list, they will all become obvious as you watch): 1) Sleeping, 2) Not participating in the debriefings except to repeat what others have already said, i.e., there is no original, experientially based material. 3) When others are actively using the gestures, they sit with hands still, or consistently make one gesture and no others. 4) They go to a great effort to consciously make the gesture in a manner that is intended only for you to see. This results in a movement so complex that it is virtually worthless as a conditioning tool. 5) While the majority of participants have become very still, they continue to fidget and move in a fully conscious manner. They often repeatedly open their eyes and shift in their chairs. 6) People who are not getting it return to normal consciousness very quickly. The others, to the contrary, fully enjoy the states and take their time in returning. 7) Some people will persist in using inappropriate exemplars despite multiple corrections and admonitions in class.

Unconscious movement is usually typified by a hesitant, almost jerky pattern of motion. It is often very slow. People who are fully conscious move differently from people who are in trance. Even if there is an isolated instance of nearly normal movement, for people in trance, that motion will emerge from, and quickly resolve into, a nearly motionless state.

Watch for changes in facial expression. bilateral symmetry increases as the face relaxes. The lips swell and the mouth will sometimes droop open. You will also notice subtle movements of the head and body. Often, a re-orientation of the head to the right (in right-handed individuals) indicates a shift from internal dialog to a felt sense of the state. For persons suffering from Parkinsonism, tremors usually abate as they enter more profound states of relaxation.

It takes some practice to learn how to see these things. You will pick them up in practice with the groups. You will also find that the subtle changes in facial expression, rate and locus of breathing, can be picked up either peripherally, or as you return to a person after looking away for a while.

I usually start the One-On-Ones with an informal chat about who and how they are and then ask something about how they like the Program. Typically I actually say something to the effect of "Well, are we doing you any good?" This is quickly followed by questions about the techniques themselves.

If the participant acknowledges no problems, I then ask him or her to show me the five gestures and the name the states associated with them. Someone who has at least been paying attention should be able to do this. I also ask them to describe the original memories that they used to create the anchors. This allows me to make sure that they are using appropriate exemplars. If they have used inappropriate exemplars, I will help them to re-anchor them later in the session.

If the participant knows the names and gestures, I begin to test how they access the states. Have them fire off one of the states. I usually choose one at random so that I have another check on whether they have actually been practicing or not. As they fire off the anchor, I look for several things. 1) Do they appear to be thinking about something or making an effort after they make the gesture? 2) Do they readjust their body before they fire off the anchor? 3) How fast do they pump the gesture? 4) Does the rate of pumping change? 5) Do their postures, facial expression and breathing style change from gesture to gesture? 6) Can they express the difference between the states?

Those participants who seem to be doing something after making the gesture, are often using the gesture as a mnemonic, not as a conditioned stimulus. Check this by stopping them shortly after the state seems to be coming on and asking them to explain how they got to the state. People who are using the anchor as a mnemonic will usually report that they go to the memory first, or they think of the picture, or they do something other than just make the gesture and notice the feeling. It is often obvious that these people have practiced the states and enjoy them. To remedy this, have them test the state. "Just blank your mind and make no effort. Put your fingers together, and as soon as you feel something in your body, but before the picture comes, pump it. As soon as you feel that, before the picture has any chance to arise, pump it. Get the rhythm of it, and pump it and pump it, and pump..." This will often solve the problem of the mnemonic. Let them stay with the state until they don't want to come back. Most participants who have had this problem report a dramatic improvement in the quality of the experience. After practicing this with a few of the states, they generally get the idea.

Some people make preparations of one kind or another before entering the states. You will see them change posture, take a deep breath and even try to get relaxed with neck rolls and stretched before making the gesture. Stop them immediately and have them perform the same kind of state test required in the previous example. Explain to them that the more preparation that they go through, the less utility the anchor has. Most of the utility of the conditioned response is in its immediate availability. So, the simpler the behavioral sequence by which it is invoked, the more effective it becomes.

Pumping styles are sometimes problematic. Many men are tempted to pump forcefully. Watch the tension in the arms and hands of the participants for excessive tension. Remind them that the gesture should be relatively effortless.

Some of these same people will fully open and close the gestures, like the pincers of a crab. This results in a flashing or weakened felt experience. Coach them through the testing part of the exercise but prod them to move faster. Tell them not to wait until the experience increases to peak (that is often their error). Tell them to pump, pulse or rub at the first hint of a response. Show them how to fire the anchor by pulsing the gesture, rather than taking the fingers apart. Emphasize the rhythm and coordination between their hands and their bodies. I sometimes use the metaphor of catching flies. The response must be quick, like catching flies. Do it so quickly that you can't wait to get the full feel of the state.

There can be changes in the rate of pumping. Some people will get so 'spaced out' that they will forget all about pumping the anchor. The rate usually begins slow and then speeds up. You often have to encourage people to pump faster. First they are coordinating the relationship between the gestures and their body. When you become aware of their bodies relaxing and the kinds of deepening signs mentioned above, begin to exhort them to pump faster. Call their attention to the fact that they can now control the intensity of the state with the speed of their pumping. Let them know, too, that as they turn their attention to some delightful facet of the experience, continued pumping will tend to center that at the heart of the experience.

From time to time you will get people who do not seem to change much as they go into state. Most people display subtle differences in physiology from state to state. Most people, when quizzed about this, can tell you something about the differences. If you see no difference, and they express no difference, be suspicious.

In some cases, this lack of difference has meant that the participant had anchored dissociated exemplars for each of the states. What was anchored, in these cases, was only a mildly interesting feeling about a series of dissociated memories. Because none of the memories were associated, all of the feelings were identical. These people need to re-anchor each of the exemplars as a fully associated experience. Take them back to each exemplar and have them notice the difference between seeing themselves as if in a movie and experiencing the memory from within their own body. Have them practice stepping fully into the experience, spinning up the feeling and re-anchoring the state. If they have been reasonably diligent with the dissociated anchor, the re-anchoring process can go quickly, and may be completed in one or two sessions.

There are also people who display the same flatness about the anchors because they are faking it. They have learned to go through the motions but talk away or strongly resist any experience. This is often a sure sign that they are using. In these cases, the clearest test is emotion. If they have the states working, even if they enter angry or upset, a short session with the anchors will change their mood. Fakers also will resist cooperation. Even if they seem to cooperate, their mood will remain unchanged and they will be unable to provide details of internal experience.

Inappropriate exemplars are, most commonly based on the participants' interpretation of the descriptive words (Focus, Solid, Good, etc.), rather than upon the paragraph that explains the state. So, Focus becomes an awareness of their need to keep focused rather than a memory of positive focus. Good is often misunderstood as something pleasant that happened, rather than an experience of revelatory understanding. Solid is often mistakenly connected to a decision that was either an emotionally driven non-decision, or something that did not include the requisite process of choice.

Bad choices may also be based on a series of often repeated events (Fun, as summers at the beach) rather than ten seconds of one specific event (that day at the beach in 1993 when...). Some bad choices represent an entire day or several hours of activity. Each exemplar must be limited to about the best ten seconds from one specific event in time.

In some cases the exemplars are all rooted in the same emotional or physical event. One recent participant linked all of his examples to his girlfriend who is now in Iraq. He felt that because she was so important to him, he would want her to be in all of his thoughts. He is also terrified that she will reject him when she returns.... Another man connected all of the examples to a specific basketball game. It is crucial that the exemplars be drawn from different contexts and that each of them connects to a context that is complete and satisfying in itself. Completeness—not subject to further modification or corruption—is a crucial part of the definition.

For people who have anchored inappropriate exemplars, it is important to re-anchor a new memory for each wrong one. It is sometimes wise to spend some time cleaning up the good ones first. This will get them into the kind of physiology that will make the change easier. It will also begin to resolve issues rooted in mood. Further, if you've gone through the enhancements listed above, they will automatically apply them to the re-anchored state.

When you get people who are uncooperative, or who refuse to participate appropriately, their expulsion (usually under threat of starting over) validates the serious nature of the Program and the value that you place on it.

For people who are doing well, I make sure that they get their pumping timed so that they can achieve maximum intensity. As each participant goes into state, I provide a certain amount of coaching using hypnotic language patterns. Remember that the deeper into state people can go and the more pleasure that they can derive from them, the more useful these states become.

Application Notes

I have attached the success criteria from the forms section of the manual. I think these notes provide a better overview, but here they are:

1) Name the five states and illustrate the appropriate hand gestures; do this in order (Exercises Two and Three).

2) Describe the sequence of physiological responses as the state arises (Exercises one-five and throughout the Program).

3) Physiological signs: Changes in posture, facial expression, heart rate, breathing and skin tone. Many will begin to express rhythmic movements that reflect the underlying experience. Although they differ from person to person, state changes will be observable.

4) Response latency: Persons who have entered the deep states required will either not respond to external stimuli (loud noises) or will respond with marked latencies (eye movements several seconds after the sound)

5) Perseveration: Persons who access the states appropriately usually take a few seconds to return to normal consciousness. Persons who immediately return to normal voice tone and reaction time are suspect.

6) Mood change: The state enhancement and anchoring exercises (one-five) and all of the others, lead to strong positive feelings. People who begin the session in negative states quickly change to more positive affects. Persons who retain a negative mood are suspect.

7) States arise automatically in response to the anchors; there is no prep time or conscious effort to access the state.

Application Note 6

All through the Program we have been working to enhance choice, self esteem and a sense of self efficacy. The anchors, as specific affective tools, have laid significant foundations for this. People who are following instructions are by now spontaneously reporting positive effects in their every-day life. Often, without specific instructions to do so, they will have begun to use them in real-life situations.

Getting to NOW brings us to a deeper level of experience and modifies the skill set by adding a deeper, more fundamental state to the repertoire. It is designed to begin a constellation of a sense of the deep Self.

I usually spend most of this session going in and out of the states. This is to make sure that everyone has them working well. I will often begin by having the participants spin up the resources and only use the anchoring gestures after they have gotten the states to ecstatic levels (Pattern Ten from Exercise Four). For those who have not mastered all of the anchors, this provides a quick opportunity to get the anchors working. In the first of these reviews, start with the anchors that work well for everyone. This gets them into state and makes the rest of the session flow more smoothly. After reviewing all of the states once or twice with this pattern and taking time to discuss how they have been using the states in the past week, I will move to the constellation of NOW.

Even at this point, there are often a few people who have not mastered all of the states, so, I will end the last review (using pattern ten) with the weakest or most underused anchor. For those who have not mastered the anchors, they at least begin with an enhanced experience of a state that they need. Also, for the others, the weakest state gets enhanced at the outset. The first several times that you do this, you will probably want to use the regular order either straight through or in reverse. Later on, when you have familiarized yourself with the states, you can vary the order.

Start with pattern ten and at the peak experience of the imagined state, have them actually fire off the anchor (e.g., start with focus). Spend some time talking them through the various dimensions of the feeling. Have them notice the internal dimensions of depth and height, breadth and extension in space. Call their attention to the differing patterns of tension and relaxation, how it moves and where it feels still. Let them notice the texture of the feeling and the patterns of warmth and coolness. Have them imagine the feeling as a single pattern; give them a few seconds to appreciate the experience as a single complex and ask them to hold this perception in mind. As they do that, as they hold the feeling in mind as a single pattern, instruct them to move their fingers to the next stimulus gesture – as they hold the previous pattern in mind — and notice how it manifests in relation to this. Spend time describing the dance between them, the complimentary and contradictory aspects. How the patterns of colors and lights of tensions and relaxations intertwine and dance. Use language that describes interweaving and intermingling, integration and synthesis. Point out the beginnings of a sense of something new that begins to arise between them. After a few minutes, as the pattern coalesces, again draw their attention to the entire pattern: to the harmony, and the pattern, the single manifestation of the interaction of the two. Instruct them that as they become aware of this pattern of integration, they should move their fingers to the next gesture and, while they hold that pattern in mind, begin to pump the gesture. Here again, call attention to the pattern of interaction as the new feeling rises up and integrates with the others. Repeat the same kind of patter with this and the follow the pattern through the remaining gestures.

When you have reached the final anchor, have them hold the combined pattern in mind. Suggest that they notice how it grows and expands. Point out that it might be both very new and quite familiar. As you do, suggest that they gently make a fist and begin to pump the gesture for the NOW anchor. Remind them that their brains already understand the meaning of anchors and pumping and to enjoy the strengthening and continuing resolution of this new state. As they continue to pump the new anchor provide further suggestions: "Allow yourself to find, in the most relaxed way, *that's right*, you can just notice, REALLY NOTICE, the deepest most empowering part of this state. Notice that it goes ALL THE WAY DOWN, Right down THERE down in the center... of the very Best part.... of WHO.... YOU REALLY ARE... and who.... you've ALWAYS BEEN." Again call their attention to the dimensions of the new state, spatial, temperature, moisture, texture, floating. Remind them to "REST..., **ALL THE WAY** down, down... into the very.... best part. And you can NOTICE,... REALLY NOTICE , the more you RELAX INTO IT, the more control you have...."

The paradox is important. By resting more completely into the state, by "drifting... and floating and LET-ting GO..." they will discover a deeper sense of control over the depth and quality of the state. Give them lots of time to enjoy the state during much of it you can repeat many of the same kinds of suggestions. Then bring them back and debrief. The same criteria apply for this as for the other states, flashing or fading requires a faster pumping style, overexertion breeds frustration, etc.

After you debrief have them spin up the NOW anchor (using the technique from Exercise Four, Number Ten) and only fire it off after it gets to an outrageous level. Let them spend time there. You may want to go through the pattern of suggestions in New Worlds to Gain. This meditation is, in principal, a 'deepener' that uses dissociation and pseudo-orientation in time to enhance the state. As a simple proposition it asks them to step into the experience that they would have of the state if they had been practicing it for ten years. It then sets up a loop between now and then and then and now. The confusion between the two states is intended to magnify the perceived enhancement.

People who have trouble getting to NOW often have an ill-formed exemplar for one of the states. It is often either tied to content, not anchored at all, carries emotional baggage or is not derived from a single, simple, memory experience. Discuss any problems and, as a first line of defense, review the qualities of the exemplars.

Assign practice generating and eliciting the NOW anchor for homework. If you have time, test the anchor before they go... sit up in a chair and make no effort other than making the fist and pumping. As you notice the feeling begin to arise in your body, at the first hint, pump. At the first suggestion of the new change, pump. Pump it and pump it and pump it...

Application Note 7

Now that NOW has been installed, one of the first things that you will want to do is go through it again. First get a sense of how well everyone did over the past week. If everyone has it working pretty well, have them spin up their best experience of NOW and when it gets much stronger than they expected, have them fire off the anchor. If several people are having trouble with the NOW state, have them review each of the anchors using Exercise Four, Technique Ten and re-anchor the state.

Bring them in and out of the state several times to ensure that they have a solid experience. After each visit to the state, I usually ask questions about their experiences. I also look to do some trouble shooting with problems. If now is unpleasant, I look for a badly chosen anchor from the initial five or a failure to lose the picture.

After several returns to NOW, I will return them to the state and use the New Worlds to Gain Meditation. It is the only one I use consistently. I have added some emphases. Needless to say, I ad lib the whole thing and as you become comfortable with the process you find yourself developing your own style. This sets up a powerful feed forward from the present experience into an anticipated future experience. This is usually the first place make use of it but I will also employ it at other times; here or there, NOW and Then, as it seems appropriate.

When we have well reviewed the NOW state and discussed the experience of the future NOW, I begin explaining the homework.

For Exercise Six, Pacing the Future, I set up a Journaling exercise. Each participant is instructed to practice NOW just as they did for the other five anchors. They are to start off each morning with the anchor and end each night with it. Upon arising, after firing off the anchor and from within the anchored state, they are to identify three places during the day where they anticipate the feeling of the anchor will be useful. They are to write these down and give them no further attention. The intent is to produce an anticipatory set for using the anchored state in those contexts. They are also required, before continuing the day, to make three appointments with themselves through the day. At each of the appointed times (they should vary from day to day) they will: 1. Fire off and enjoy the anchor (eyes open or closed, privately or publicly). 2. Take a few minutes to review how well the day has gone to this point. At this time they will make a few notes on their accomplishments and then get on with business. At the end of the day they will again fire off the anchor, review the day and make some progress notes. Finally they are instructed to drift off to sleep firing off NOW.

Once again, the intent here is to create connections, to seed generalization of the positive resources throughout their days. This overcomes the problem of having a skill that is limited to the practice hall but is otherwise unavailable. It also provides an opportunity for them to reflect on the value of the techniques in real life.

I always warn them that this exercise will be collected. Sometimes, when I cannot be present, I will extend the exercise for two weeks and have them send in the first week's exercises. Most people enjoy the exercise and return to the sessions with interesting tales..

End the session with an experience of NOW (if there is time).

Application Note 8

Exercise seven is pretty straightforward. The participants are to create their own anchors. They choose the resource, real or imagined, they choose the gesture and they are responsible for cranking it up. Once they have made and tested the anchors, have them use the techniques from Exercise Five (Getting to Now) to integrate these into NOW.

I always take some time to emphasize that this is an important point in the program. This is where control is officially returned to the participants. By creating their own anchor, they come to know, in no uncertain terms, that this is their skill, operable on their own terms. Spend some time in the next session reviewing the anchors that they have created and making sure, by questioning, that they have actually anchored them and that they have appropriately specified the root experience. Here too, it is crucial that the state anchored has been thoroughly dissociated from content and context.

It must again be emphasized that all of these anchors must be positive states that make you feel good. I recently had an experience where a participant had anchored his feelings of longing for his family and retained content and context. As he used the anchors he grew more depressed and frustrated over his inability to be with them. This ended in a minor relapse.

Throughout the program it will be important to insure that all anchors are content free. We start with the memory to get to the feeling. We then enhance the feeling, independent of context before anchoring it to the stimulus gesture.

One of the interesting things about Exercise Seven is that participants often find that their own anchors are stronger than the ones I've specified. This, of course, just goes on to emphasize that the skill is theirs.

To begin the session, encourage the participants to compare notes about the exercise. Poll the group to find out what they anchored, how easy they found the process and whether they find their anchors stronger or weaker than the first five. As you discuss the anchors get specifics. Find out the initial memory, the physical gesture and how they experienced it. Do not allow anyone to get by with "It feels good ...." Find out if they all got it to the float, without pictures and have anchored it so that it is totally automatic.

After some discussion, ask them "What do you know this week that you didn't know last week?"

The answer that I'm looking for is that they can do it without our aid. All of these experiences are uniquely theirs and have nothing to do with probation, facilitators or anything else. It is all theirs and part of our plan to return themselves (gift themselves) back to themselves. Let them know that this is important.

Have them access the new anchors one-at-a-time by first remembering their best experience of the anchor, spinning up the feeling and then, once it's become strong, firing off the anchor and pumping (see Exercise Four, Number Ten). Give them time and these basic instructions for each of the new anchors. Do it once for NOW as well.

Once they have recovered......

Tell the group that you are going to show them how to integrate their own anchors into the NOW anchor. Tell them that we will begin with NOW and then add in the anchors that they have constructed. Also let them know that the process itself with enhance their own anchors further.

Here is the pattern of the following meditation:

    1) Fire off NOW.

    2) Focus on the feeling of NOW as you fire new anchor.

    3) Become aware of integration.

    4) Hold integrated feeling in mind.

    5) Fire off NOW..

    6) Loop through pattern until all exemplars are integrated.

    7) End with NOW.

Instruct the participants to fire off the NOW anchor and pump it up until it is really strong. Wait a few minutes, with or without suggestions. When *that stillness* settles over the room begin....

That's Right....

NOWWWWW........

Allow your attention

to gently

settle down

into the very best part.

ALL... the way down.

And as you do....

As you pump it … and pump it … and pump it...

Allow your attention to focus on the feeling and the Pattern of movement

as the feeling moves through your body.

Hold it in mind … and body … as a single feeling … a single perception.

And as you do... As you focus... on the very best part of the feeling...

Hold it … in mind and body   as you gently

fire off the first of your own anchors.

That's right,

fire off the new anchor   as you keep your attention

focused

on the feeling of NOW....

And as you pump it … and pump it   … and pump it

you can become   … aware   of that new feeling   rising up in the background

awakening new perceptions,   New possibilities   New dimensions of NOW.

Become aware … of how   … it becomes   the center … of

NOW ...

Opening … New dimensions, …New landscapes

and as you pump it … and pump it … and pump it

JUST … REST …down   … into the very best part
of this new feeling

And hold it …in mind   … and body

Notice how it moves

Notice how deep it goes

and as you focus   down ….into the very center

Hold that feeling …   in mind and body   …and gently   … move your fingers

back to NOW … and pump it …and pump it and pump it.   …And as you do....

As you pump …. NOW …..

You can notice... … Really notice...

New dimensions....... New perceptions....... New landscapes....

NOW.

Allow your attention

to gently …. SETTLE …   down

into the very best part.

ALL
the way
down.

And as you do.... As you really do...

As you pump it … and pump it … and pump it...

Allow your attention … to focus … really focus

on the feeling … and the pattern … of the movement

as the feeling

moves

through your body.

Hold it in mind … and body

a single feeling … a single perception.

And as you do…

As you focus… … on the very best part …

***Hold it in mind and body

as you gently … fire off the next … of your new anchors.

That's right,

fire off the new anchor … and keep your attention … focused… on the feeling

NOW….

And as you pump it … and pump it … and pump it

you can become aware … of that new feeling rising up in the background

awakening new perceptions, … New possibilities … New dimensions of NOW.

Become aware … of how it becomes … the center of NOW…..

Opening New dimensions, … New landscapes

and as you pump it. … and pump it … and pump it

JUST REST down … …into the **very** best part

of this new feeling

And hold it … in mind and body

Notice how

it moves

Notice

how deep

it goes

and as you FOCUS

DOWN … into the center

Hold that feeling … in mind and body

and gently … move your fingers … back to NOW

**....continue similarly with the third, fourth and fifth anchors...

Here is the pattern again:
1) Fire off NOW.

2) Focus on the feeling of NOW as you fire new anchor.

3) Become aware of integration.

4) Hold integrated feeling in mind.

5) Fire off NOW..

6) Loop through pattern until all exemplars are integrated.

7) End with NOW.

You may want to go through it again if you have time. If not, just have them fire off NOW and pump it up. Let them enjoy it for as long as you can. Erickson thought nothing of leaving people in trance for 20 or 30 minutes.

Application Note 9

Exercises Eight and Nine are the original heart of the program. They are the oldest stratum and represent my original thinking about the generative change that is so important to the program. The resources that they evoke are in some ways less practical than the first five anchors. These are designed specifically to awaken a positive sense of Self. They were also chosen to provide a dynamic sense of personal direction from within. I originally called the exercise and the state that it creates the *Feeling Toned Vector*. The name implies the felt sense of direction that the state awakens and refers back to the Jungian idea that ideas are linked into complex factors or complexes by similar feeling tones.

Another important aim of the exercise is the awakening of the positive past. Most people in life have left behind them a great deal of positive experience that can contribute significantly to their development into the future. This is especially true of persons who have been subject to imprisonment and addiction. They have often lost all contact with their positive past. Here we use state dependent memory effects to awaken a more general identification with all that's best in the participants.

The exercise is relatively simple. Participants are to provide a series of memories that exemplify the categories. There are a few wrinkles.

In the first category I prefer fantasy aspirations, like dreaming about being superman, or a fireman. It is important that the end result for each of the aspirations is a felt sense of excitement and anticipation. The examples here should never result in a feeling of loss or missed opportunity. This is an important reason to go for the fantasies. Have them remember a specific example of a time when they were dreaming. Make sure they get fully into it.

More generally, assuming that the participants have learned to enhance an anchor state effectively, the requirement that the exemplars must be clean, legal and sober no longer holds. Insofar as we have developed the skill of abstracting the feeling from the memory (and by combining them later we will further dilute the content), these considerations are no longer of primary importance. In the beginning they were important for practical and political reasons. The practical reason is that during the first weeks, many participants hang on to enough content that it can get in the way. By now content should be moot. All participants should be able to go directly to the feeling. The political issue should be obvious. It would be unseemly to have felons fantasizing about their hopes for illegal livelihoods early in the program. Even if we assume that the skills can be ramped up quickly enough, it just wouldn't look right. Now, however, we can appeal to the learned skill set as insulation against the possible negative effects of criminal aspirations. Nevertheless I don't encourage the use of negative exemplars.

Throughout this exercise, as in any other, it is crucial to emphasize that each memory MUST be of a single specific event. As they share the memories in class, challenge each person to provide specifics. Be especially aware of those persons who talk vaguely about this or that. Nail them down to time and place and get specifics. All of these exercises depend on a high degree of sensory specificity.

When people have difficulty creating a strong anchor here, the most common problem is an ill specified exemplar. Nine times out of ten, I've found that weak anchors are the result of vague, non-specified exemplars.

Jobs and Roles should be fairly obvious, but participants need to be reminded that these MUST be examples that made them feel good. Things you do well can be any skill or ability that you have mastered and do with ease. It does not matter if it was easy to learn or difficult. It must be something that fills you with confidence or provides a sense of ease. Suggest that they remember a specific incident when they appreciated the skill. It might even be a present time appreciation of the ability. Things You Learned Easily requires the element of ease. Whether it was worth learning or not doesn't really matter. Some things are learned easily as a matter of innate disposition, others are learned through appropriate motivation. Both are acceptable. Their continuing value or utility is unimportant.

Times You Really Felt Good About Yourself is a place where I encourage people to add in accomplishments, reveries, feelings of wellness, health, spiritual experiences and mystical experiences. They should all be more about 'me' than about 'what I did' but that is sometimes more difficult than it appears.

When the class reconvenes, I generally start by asking for examples from each category. In more difficult groups, I poll the participants one at a time for each category. It is my hope that they will get excited and that one batch of memories will excite others in their fellows. Generally, this session tends to degenerate in a pleasant free-for-all in which the participants share their recollections with one another. When this works you just need to provide sufficient guidance to make sure that all of the categories get covered and that there is enough time to anchor the states. As this goes on, you'll usually find that people who missed the last session will have ample opportunity to find their own examples and they can take notes during the session. Encourage them to do so. You can also encourage some of the others to add new inspirations to their lists.

Application Notes

When you have determined that everyone has chosen appropriate exemplars and they have all spent some time discussing them, then instruct them in the review and anchoring phase. I generally go through the anchor categories one at a time and have them spin them up before anchoring.

Begin with *Things You Wanted To Be*. Instruct them to get the first example, "or your favorite". Step all of the way into it and enhance it, spin it, until the content disappears and you find yourself floating. Hold onto that feeling and add in the next example from the same category. Spin that into the mix until its content disappears. Then add the next. Add all of the exemplars from the category and have them float for a while, enjoying the new state. Bring them back and briefly discuss the new feeling. Repeat for each of the categories.

Two things are to be noted: 1) the process should work much more quickly than the initial efforts at anchoring and enhancement and 2) because we are layering in multiple images and contexts, there should be little problem with content and contexts; they should cancel each other out.

When you have reviewed each of the new resource states, start to anchor them. Follow the same process but when they reach the **combined** float state, have them tap the appropriate finger. The manual says to start anchoring with the first exemplar. I no longer think that this is necessary. Let them combine them, get to the float, and then anchor the mixed state.

The anchoring process changes here too. Rather than bringing them in and out of the state, have them begin tapping when they get to the float. Spin up the feeling and as they feel it increase, tap again. Spin it up some more, tap again. Continue until it catches, then tap just as we pumped the others. Shake it out. Test it and go on to the next.

I usually start off by talking them through the first two and have them do the rest at their own speed. If time gets short, lead them through, reminding them how good they are at this skill.

When all five have been anchored you can combine them into NOW. Alternatively, you can invite them to practice the new anchors until the next session and to work at integrating them into NOW on their own.

If you decide to do the NOW integration start by having them fire off the first anchor (Thumb) "Thump your thumb, and as you begin to feel the feeling of the anchor starting to rise up in your body, thump again. As soon as you begin to notice that, thump again..." etc. Suggest that they notice how the state arises in the body, where it starts, how it spreads, its temperature, etc. Instruct them to pump it up and pay attention to its feel, where it centers, its patterns of relaxation and tension, its warmth, its character. Suggest that they hold the felt pattern of these perceptions in mind as they move their fingers into the NOW gesture. Let them continue to focus on the felt sense of the original state as they pump NOW and note how NOW arises and integrates with it. As they become aware of the dance between the two states, pump NOW and enjoy their integration. "Notice how this state opens new dimensions in the NOW state. As NOW gets stronger, hold its new pattern and character in mind as you gently release your fist and begin to thump your index finger..." etc.

Move through all of the five new anchors in the same manner. When they get through all of them, you may wish to either return them to the excitement of what you wanted to be, or just leave them enjoying and exploring the new qualities of NOW. Give them ten minutes or so of uninterrupted exploration and then bring them back and debrief. If there is time, have them fire off NOW and spend a little more time with its new features.

General Session Procedures

After the first session, all of those that follow are organized according to a standard outline: Greeting, Homework Discussion / Problem Resolution, Review of New Material, Homework Assignment, (optional) Meditation and Dismissal.

Greeting: At this time the participants are welcomed. Issues that may exist outside of the session are dealt with. Procedural problems are addressed and attendance is taken.

Homework Discussion / Problem Resolution: Before any new material is introduced, the homework and the previous week's material are reviewed. Problems with interpretation of the exercise are discussed and clarifications made. Where necessary, the previous lesson may be reviewed in its entirety. All questions should be addressed in a respectful manner and answers to questions given as full attention as possible. The general aims of the exercises and the logic of the program may be discussed. The emphasis should always be positive.

Each lesson carries with it presuppositions and behavioral outcomes that are specified for each. The instructor should be aware of them. Persons who are unable to meet the behavioral criteria may need extra attention.

There will be times when participants need to vent or discuss problems in an open forum. This is the place to do it, but it should not be allowed to supersede the stated intent of the sessions which is to teach cognitive/emotional skills. Persons having serious problems with the material may need to be seen in one-on-one sessions. Troublemakers should be asked to wait until after the session.

Review of New Material: Once it is clear that everyone has understood and completed the work on the previous material, new material can be introduced. Follow the instructions in the exercise and refer to the instructional notes. Make sure the emphasis is on getting results. Our results are strong experiences of positive states that are increasingly under the control of the participants. One cannot underestimate the impact of the depth of the experiences. The more profound the experiences, the more lasting and useful the changes will be. This is not a program from which one can profit by merely listening. The exercises **must** be practiced. In most contexts every participant can be led through the exercise to some kind of positive result. Take the time to talk as many as may need it through the experience. **Be demanding**. Accept nothing less than answers in precise sensory terms. Use every skill available to create deep altered states of consciousness.

Homework Assignment: Every session has a homework assignment. In general, it is an assignment to practice the exercise introduced and/or to complete it before the next session. Persons who consistently fail to do the homework should be invited back at a less convenient time to review the material. An important aspect of the homework assignment is its capacity to separate the behavioral skills from the treatment context. By succeeding at the homework assignment away from the treatment setting, the participants gain a sense of self-efficacy that lies at the heart of the program. Further, homework seeds generalization of the affective skills taught here to other contexts.

Meditation: Each session may be brought to a close with a meditation that is keyed to the content of the lesson. These experiences are not crucial but often add to the quality of the experience as a whole and can serve as a reinforcing break at the end of a long session. All of the meditations are scripted. Basic instructions on the use of hypnotic language and intonation are provided in the introduction to that section. Participants often express a great deal of positive anticipation towards the meditations. All of the scripted meditations may be provided to the participants as tapes or CD audio files.

Dismissal: Although not a formal part of the class it is important to ensure that all business has been completed. Participants will not be dismissed until they have remained sufficiently attentive that all of the session's business can be cleared. The sessions should end on time. From time to time there may be pressing issues that monopolize the group's time. If this happens, acknowledge it and ask the participants to provide the level of respect to the participant that they would like for themselves. Before dismissing the group ensure that those persons who need to stay later have been identified.

Urinalysis: If urinalysis is required for participants and it is not collected at another time or place, schedule the specimen submission for times before or after the session. Do not permit problem urine submissions to interfere with the schedule for other participants. If participants are late for specimen submissions, let them wait. Since the sessions run about two hours, most participants should have no problem in waiting.

Program Timeline

In general, the program as presently implemented follows the following sequence. Needless to say, changes can be made. Two of the original exercises , Four and Nine, are now considered optional in favor of more practice. In general, when in doubt, spend time practicing and spending time in the altered states.

Approximate Timeline for the Brooklyn Program. Use more or less time as needed. Times for each exercise will vary with the group.		
Week	Exercise	
1	Introduction & Exercise One	
2	Exercise One continued (several examples)	
3	Exercise Two (one or two states)	
4	Exercise Two continued (two or three states)	
5	Complete Exercise Two begin Exercise Three	
6	Continue Exercise Three	First One-on-one
7	Complete Exercise Three begin Exercise Five	
8	Review Exercise five	
9	Exercise Six	
10	Exercise Seven	
11	Exercise Eight	
12	Exercise Ten	
13	Continue Exercise Ten	Second One-on-one
14	Exercise Eleven	
15	Exercise Twelve	
16	Recap, re-evaluaton and review.	

The One-On-One Sessions

In general, the Program requires a minimum of two one-on-one sessions. These two sessions are used to assess progress in the program and to ensure that the participants' other needs are being met.

At other times one-on-one meetings are arranged for make-up sessions and to address incidents of drug abuse while attending the program. In the case of make-up sessions, the session flows from the nature of the material missed. It should not dwell on guilt trips or threats; just insuring that the participant has completed the assigned homework and that they understand the material that they missed. These sessions also provide a good opportunity to assess current progress.

Behavioral Criteria

At every level, The Brooklyn Program defines verifiable indicia of the level of cooperation provided by the participants. Because it is rooted in conditioned response systems, everyone who provides even the most rudimentary participation will manifest state changes that are predictable and reliable reflections of participation. In fact, the elicitation of the states themselves serves both as a tool for installation and testing of the level of cooperation. Those who are not engaged can be identified by their physiological responses.

In the early stages of the Program, many participants attempt to slide through without practicing the states. By providing in-group practice and intense one-on-one installations (all of the states can be installed in a one-hour one-on-one session), the level of compliance is measurable and almost guaranteed. Once over the hump of mastering the first five states, compliance is relatively assured. There are, however, always holdouts.

In the process of conditioning the first five states several indicia are presented.

1) Most simply, the participants who are at least paying attention will be able to name the five states and illustrate the appropriate hand gestures. They will be able to do this in order.

2) Physiological signs: Participants who enter the states will show profound changes in posture, facial expression, heart rate, breathing and skin tone. Many will begin to express rhythmic movements that reflect the underlying experience. Although they differ from person to person, state changes will be observable.

3) The exercises require the participants to describe the sequence of physiological responses as the state increases. Although participants may find it difficult at first to describe the feelings, non-participants will be unable to define any sequence.

4) The anchoring exercises, almost without exception, lead to strong positive feelings.

5) Persons who have mastered the conditioning exercises will experience immediate mood changes when they make the anchoring gestures. If they come to the session in a bad mood, the mood will not last.

6) Persons who have mastered the anchoring skill will need no preparation time before accessing the state; the states will arise automatically after making the gesture.

7) Persons who have practiced the states will become generally unresponsive to external stimuli. When they respond, they will exhibit slow reaction times and decreased orienting effects. Most people who are enjoying the states take some time to reorient after accessing the states.

A full set of these criteria is provided in the forms section.

It is important to differentiate between persons who have learned the process incorrectly from those who have not made the effort. A typical mistake made in the anchoring exercise is a tendency to continue using the memory image to access the feeling, long after the response should have become an automatic felt response. Participants who make this error often close their eyes and seem to be spending some time in preparation before making the anchor gesture. When asked, these participants will usually tell you that they are going to the image to get the feeling.

Instruct persons who make this error to do the following:

1) Close your eyes. Blank your mind and make the gesture. Make no effort to remember, **just notice** what happens in your body when you make the gesture.

2) When you feel something happening in your body, make the gesture again and **just notice** how the feeling changes.

3) As soon as you notice another change, make the gesture again.

4) Continue in this manner until the feeling becomes very powerful.

5) As they now pump the gesture, provide suggestions that they notice the movement of the feeling, how it grows, whether it has temperature, if it has texture. Take care to suggest that they notice sufficient detail about the feeling that the other senses are completely overwhelmed. Lead them through the same process with each anchor and reconstruct the NOW state.

Like the initial states, exercises focused on either creating individual resources or making resources available in other contexts (future pacing) provide similar indicia to the root states. Dissimulators will be unable to describe the sequencing of sensory data as the states develop; they will fail to reflect the appropriate postural and physiological indicia of state change.

In the positive resources exercise, each individual participates in a discussion of the significant life events that they have chosen. Dissimulations are usually obvious and reveal themselves in the following manner.

1) The initial segments of the exercise are done in writing, giving clear evidence of cooperation on that level.

2) Sometimes, the participant claims to have no positive memories. The excuse is most often a cover for non-preparedness. When it is honestly made, it is usually a reflection that the participant has failed to open access to the past by not participating in the anchoring exercises.

3) Group response. Those who have participated honestly will often provide powerful feedback to perceived liars.

4) Lack of pattern. Part of this exercise is the elicitation of patterns of competence and personal direction. If the facilitator opts to focus on a discussion of the underlying themes and no unifying patterns arise, or if the patterns are blatantly superficial, it is usually a reflection of non-cooperation.

For *Positive Resources Revisited*, the anchored state that results is usually experienced as strongly positive while differing in individual characteristics for each participant. The non-compliant individuals fail to show the clear changes in physiology and unconscious indicia of pleasure and pride. More often than not, the participants express spontaneous surprise at the end of the exercise.

Smart Outcomes is a highly interactive exercise. Persons who actively apply the lessons gleaned from the earlier exercises emerge with a powerful felt sense of future orientation and motivation towards the outcomes expressed. The non-compliant typically produce an ill-formed, passive, already perfected, or uninteresting outcome. They can often be led through the process despite themselves. The Smart Outcome exercise is a point in the program where some persons who have been otherwise disengaged awaken to the value of the program.

One of the important indicia of success or failure in the Smart outcome exercise is the nature of the outcome sought. Superficial outcomes, cars, money, fame, are typically evidence of non-compliance or superficial compliance. The outcomes for this issue should flow from the deep experience of NOW. They should be about what the participants will be doing, not what they have. Require the participants to describe their activities not their possessions. In those cases where the responses are superficial; test the resource states and seek to lead them to a deeper experience of NOW and the future that grows out of it.

Dealing with Positive Urinalysis or Slips

Since the Brooklyn Program is presented as a regimen for the treatment of substance abuse, it is inevitable that in the course of the program there will be incidents when participants return positive urines or admit to a slip. In every case we follow the advice of Prochaska, Norcross and Di Clemente and seek to frame any such isolated incident as a slip and not as a relapse. In these cases, the emphasis is on regaining stable sobriety, feeling good about stopping and discovering how to deal with the issue of temptation in the future. If the participant indicates that they are in full blown relapse it may be more useful to refer them to an appropriate inpatient facility until they have stabilized (Prochaska et al., 1994; Miller at al., 1995; Shattuck, 1994).

For those who are appropriate in their discussions of slips, these sessions can become an excellent time to discuss the direct application of the techniques and resource states already learned to the problem at hand. One such approach for participants who have not yet mastered the anchoring skill set, would be to ask how they would need to feel or what they would need to do in order for this problem to be a non-problem. As the participant defines the state, slowly assist her in refining its sensory characteristics. When have you felt this way? If you have never felt that, have you imagined it? If you have imagined it, what do you think it would feel like? Where would you feel it first? How would you become aware of it? How would it develop? Would there be motion or sound? What would you look like if you felt that way? If you saw someone with that feeling or quality how would they look? How would they hold themselves? Through these and similar questions, you can induce a state, perhaps one that the participant has never felt before. Then, using the standard anchoring techniques in the first five exercises, anchor it, enhance it, and future pace it to the appropriate contexts. End the session with the Future Perfect meditation and have them imagine spreading the resource into all of the imagined future contexts.

Sometimes a slip is a simple matter of the application of common sense rules about people, places and things. Sometimes it is rooted in a failure to respond to a real psychic, spiritual or physical need. No matter what the cause of the slip, use the available tools to empower the participant. Let them leave the session feeling good about THEIR decision to end the problem and continue with the program. In every case, use some anchor or meditation to ensure that the participant leaves feeling much better than when they entered.

One of the more useful techniques is to invite the participant to review NOW or the other states. Take sufficient time to assist them into a deep experience of the state. When they have developed a deep felt experience, invite them to view the problem situation from there, holding the state stable. Ask them: "When you feel this way, is the situation different? Feeling like this, do you make different choices?" After a brief discussion, re-access the state, enhance it further, debrief and end the session. Always ensure that the participant feels better when they leave than when they entered.

The Positive Resource Day Planner

In cases where relapse is a continuing problem, the positive resource day planner can be very effective. It makes use of the program resources but also requires the participant to consciously interrupt her day to access the resource state and evaluate progress towards doing well.

Persons who pray or meditate often report how those spiritual practices provide benefits that far outweigh the few moments that they cost in the morning and evening. This exercise relies on the same quality of temporal redemption. By taking time out to feel good, participants will discover that they work and play more efficiently. It is an effective stress buster.

The technique makes use of pattern interruption (Cade and O'Hanlon, 1993; Grinder and Bandler, 1975; Watzlawick, Weakland & Fisch, 1994) to interrupt the progress of the day with experiences of feeling good. This allows the participant to reorient their attention towards more positive ends. It further requires them to reset the affective tone of the day on a scheduled basis, making use of state dependent memory effects (Rossi, 1986; Rossi & Cheek, 1995).

In many cases where a participant will be unable to attend the weekly sessions or if the sessions must be cancelled, the Day Planner provides a powerful means for providing behavioral continuity. As a written journal, it enforces practice of the resource states and seeds generalization of positive feeling through the regular day's activities.

In practice it is crucial to emphasize that the scheduled evaluations are to be positive. Enter the state and note **how well** you did. In the evening it is important to plan to feel good for the morning and to finish the day in a positive state.

There is often a certain level of resistance to taking the time required. Remind the participants that these are very short breaks and can be done anywhere. They should also understand that it is a means of caring for one's self as valid as a coffee or cigarette break is for those who employ them. Reemphasize the value of caring for yourself. "What have you done for yourself lately?"

Positive Resource Day Planner

Upon arising ask yourself: How do I want to feel today? State it positively (as what you want) and remember that it must be under your personal control. Make it something that you feel from within, not dependent upon your external performance, other people or events. Use one of the anchored states.

How will you feel? What will you feel in your heart? What qualities will the feeling have? What kind of posture will you have? What patterns of muscular tension and relaxation will you feel in your face and neck and back and stomach and legs? How does the state flow through your being? What happens first, second, third? What will your voice sound like? What will you look like? Imagine the state fully and step all of the way into it. Get a sense of how it will feel. Really try it on; anchor the state.

Describe the state here. Use sensory specific language:

Schedule a minimum of three breaks during the day when you will stop, access the state and appreciate your progress towards it during the day. Schedule them in the following or a similar time frame: Write them in. **For each break make a note on how good it feels and how it relates to your goal state.**

DATE:_____
Time

_____ _____

_____ _____

_____ _____

_____ _____

In the evening, before going to bed, access the state. Compare it to the state that you planned. Notice how good it feels and the progress that you have made and make note of what you might like to add to make it even better. Emphasize adding to the state or trying something new. From this state, plan a state that you might like to access tomorrow.

Write down your observations here and on the flip side if you need the space:

Finish the night drifting off to sleep as you enjoy the state.

The Final Session

The final session is devoted to a review and discussion of the Program. Participants are encouraged to discuss what they found valuable, what worked for them and to ask any lingering questions.

During the course of the last session the relevance of the Program to substance abuse treatment will normally be discussed and applications to sober living made more explicit. All participants should be encouraged to participate.

Evaluations are provided and certificates distributed.

Celebrations are encouraged.

Appendix

Forms and Handouts

Positive Resource Day Planner

Upon arising, ask yourself: How do I want to feel today? State it positively (as what you want) and remember that it must be under your personal control. Make it something that you feel from within, not dependent upon your external performance, other people or events. Use one of the anchored states.

How will you feel? What will you feel in your heart? What qualities will the feeling have? What kind of posture will you have? What patterns of muscular tension and relaxation will you feel in your face and neck and back and stomach and legs? How does the state flow through your being? What happens first, second, third? What will your voice sound like? What will you look like? Imagine the state fully and step all of the way into it. Get a sense of how it will feel. Really try it on. Anchor the state.
Describe the state here. Use sensory specific language:

Schedule a minimum of three breaks during the day when you will stop, access the state and appreciate your progress towards it during the day. Schedule them in the following or a similar time frame: Write them in.
For each break make a note on how good it feels and how it relates to your goal state.

Time

____   _____

____   _____

____   _____

____   _____

In the evening, before going to bed, access the state. Compare it to the state that you planned. Notice how good it feels and the progress that you have made and make note of what you might like to add to make it even better. Emphasize adding to the state or trying something new. From this state, plan a state that you might like to access tomorrow.
Write down your observations here and on the flip side if you need the space:

Finish the night drifting off to sleep as you enjoy the state

Outcome Worksheet

1. Is it stated in the positive, or can it be stated in the positive? State it.

2. Is it under your personal control? How?

3. Can you specify three different ways in which you will know that you've gotten it if you get it?
   A)
   B)
   C)

4. Do you want this all the time? Is it appropriate everywhere? Should it be limited to a specific context?

   When do you want it and when don't you want it?

   When is it right?

   When is it wrong?

5. What will it change in your life and in the lives of the people around you?
   Be specific:

6. Experience now, in your imagination, how you will look and feel, what you will see and hear when this is a reality.
   Describe what you see and hear and feel. Who is there? What is it like? Be there Now.

7. Move backwards from the final realization of the goal to discover the steps that make it possible. List the steps

8. Enumerate five steps necessary to get from here to there.
   1).
   2).
   3).
   4).
   5).

Behavioral Criteria

The program depends upon the learned capacity to create automated response systems for mood change and behavioral control. Persons who do not respond in accordance with the following guidelines have not mastered the core competencies.

In early stages of the program participants may legitimately have trouble creating truly automatic anchors. If, at the time of the first one-on-one the participant meets the first five criteria but stumbles at the sixth, she may need coaching on the anchoring process. Attempt to coach the participant through the states, enhance them and re-anchor NOW. Make appointments for further one-on-one contacts if necessary.

If, by the first one-on-one session, a participant cannot meet the first three criteria there is reason to believe that they are not participating. If they remain unable to meet the root criteria after the second half of the program they should be asked to repeat the program or to seek another form of treatment.

1) Name the five states and illustrate the appropriate hand gestures; do this in order (Exercises Two and Three).

2) Describe the sequence of physiological responses as the state arises (Exercises one-five and throughout the program).

3) Physiological signs: Changes in posture, facial expression, heart rate, breathing and skin tone. Many will begin to express rhythmic movements that reflect the underlying experience. Although they differ from person to person, state changes will be observable.

4) Response latency: Persons who have entered the deep states required will either not respond to external stimuli (loud noises) or will respond with a marked latencies (eye movements several seconds after the sound).

5) Perseveration: Persons who access the states appropriately usually take a few seconds to return to normal consciousness. Persons who immediately return to normal voice tone and reaction time are suspect.

6) Mood change: The state enhancement and anchoring exercises (one-five) and all of the others, lead to strong positive feelings. People who begin the session in negative states quickly change to more positive affects. Persons who retain a negative mood are suspect.

7) States arise automatically in response to the anchors; there is no prep time or conscious effort to access the state.

Forms and Handouts

EVALUATION

Name: Date:

1) Name three things that you learned from the group.

2) What three things discussed in the group were most valuable to you?

3) What three things would you change in the structure or running of the group?

4) What kinds of things would you like to have increased in the exercises?

5) What exercises were least useful?

6) What exercises were most effective?

7) Other Comments?

References

Andreas, C., & Andreas, S. (1989). *Heart of the mind.* Moab, UT: Real People Press.

Andreas, S., & Andreas, C. (1987). *Change your mind-- and keep the change.* Moab, UT: Real People Press.

Andreas, S., & Andreas, C. *(2002).* Resolving shame. *Anchor Point. March, 2002, p. 17.*

Baffa, C. (1997). IQ, hypnosis and genius. http://Carmine.net/geni/geni0001.htm

Bandler, R. & MacDonald, W., (1987). An insider's guide to submodalities. Cupertino, CA: Meta Publications.

Bandler, R., & Grinder, J. (1975). *The Structure of magic I.* Cupertino, Calif.: Science and Behavior Books.

Bandler, R., & Grinder, J. (1975b). *Patterns in the hypnotic techniques of Milton H. Erickson, MD, Volume 1.* Cupertino, CA: Meta Publications.

Bandler, R., & Grinder, J. (1979). *Frogs into princes.* Moab, Ut: Real People Press.

Bandler, R., & Grinder, J. (1982). *Reframing: Neuro-Linguistic Programming and the transformation of meaning.* Moab, UT: Real People Press.

Bandler, R. (1985). *Using your brain for a change.* Moab,UT.: Real People Press.

Bandler, R. (1993). *Time for a change.* Capitola, CA: Meta Publications.

Bandura, A. (1997). *Self-Efficacy: The exercise of control.* NY: Freeman.

Bateson, G. (1979). *Mind and nature: A necessary unity.* NY: Bantam.

Bechara, A., Damasio, H., & Damasio, A. R. (2000). Emotion, decision making and the orbitofrontal cortex. *Cerebral Cortex, 10*(3), 295-307

Bechara, A., Damasio, H., Damasio, A., & Lee, G. (1999). Different contributions of the human amygdala and ventromedial prefrontal cortex. *The Journal of Neuroscience, 19(13).* 5473-5481.

Bertalanffy, L. von. (1968). *General system theory.* NY: George Braziller.

Blanchard, R., Blanchard, D., Takahashi, T., & Kelley, M. (1977). Attack and defensive behavior in the albino rat. *Animal Behaviour, 25*: 622-634.

Bodenhammer, B. G., & Hall, L. M. (1998). *The user's manual for the brain: The complete manual for neuro-linguistic programming practitioner certification.* Institute of Neuro Semantics.

Bodhi, B. (1995). *A comprehensive manual of Abhidhamma: The Abhidhammattha Sangaha of Acariya Anuruddha.* Kandy, Sri Lanka: Buddhist Publication Society,

Bouton, M. E., Moody, E. W. (Nov, 2004). Memory processes in classical conditioning. *Neuroscience & Biobehavioral Reviews, 28(7),* 663-674.

Brooks, M. (1989). *Instant Rapport.* NY: Warner.

Brown, S. M., Manuck, S. B., Flory, J. D., & Hariri, A. R. (2006). Neural basis of individual differences in impulsivity: Contributions of corticolimbic circuits for behavioral arousal and control. *Emotion 6*(2): 239-245.

Cade, B., & O'Hanlon, W. H. (1993). *A brief guide to brief therapy.* NY: W.W. Norton.

Campbell, J. (1988). *The power of myth.* NY: Doubleday.

Casteneda, C. (1993). *The art of dreaming.* NY: Harper Collins.

Chambers, R. A., Bickel, W. K., & Potenza, M. N. (2007). A scale-free systems theory of motivation and addiction. *Neuroscience & Biobehavioral Reviews, 31*(7): 1017-1045.

Cialdini, R. B. (1993). *Influence: The psychology of persuasion.* NY: Quill Publishers.

Codispoti, M., & De Cesarei, A. (2007). Arousal and attention: Picture size and emotional reactions. *Psychophysiology, 44*, 680–686.

Craig, A. D. (2009). How do you feel--now? The anterior insula and human awareness. *Nature Reviews Neuroscience 10*(1): 59-70.

Damasio, A. R. (1999). *The feeling of what happens: Body and emotion in the making of consciousness.* NY: Harcourt.

Damasio, A. R., Grabowski, T. J., Bechara, A., Damasio, H., Ponto, L. L. B., Parcizi, J., & Hichwa, R. J. (2000). Subcortical and cortical brain activity during the feeling of self-generated emotions. *Nature Neuroscience, 3(10),* (October, 2000).

Damasio, A. R. (1994). *Descartes' error.* NY: G. P. Putnam.

Davidson, R. J. (2004). Well-being and affective style: neural substrates and biobehavioural correlates. *Philosophical Transactions Of The Royal Society Of London. Series B, Biological Sciences* 359(1449): 1395-1411.

DeShazer, Steve (1994). *Words were originally magic.* NY: W. W. Norton.

De Cesarei A. & Codispoti M. (2008). Fuzzy picture processing: Effects of size reduction and blurring on emotional processing. *Emotion, 8*(3), June 2008, 352-363.

De Cesarei, A., & Codispoti, M. (2006). When does size not matter? Effects of stimulus size on affective modulation. *Psychophysiology, 43*,207–215.

Dilts, R. (1993). *Changing belief systems with NLP.* Cupertino, CA: Meta Publications.

Dilts, R. (1995). *Strategies of genius (vol. 3).* Cupertino CA: Meta Publications.

Dilts, R., & Delozier, J. (2000). *Encyclopedia of systemic neuro-linguistic programming and nlp new coding.* Scotts Valley, CA: NLP University Press.

Dilts, R., Delozier, J., Bandler, R., & Grinder, J. (1980). *NLP. vol.1.* Capitola, CA: Meta Publications.

Downs, T. H., Creem, S. H., Wraga, M., Harrington, G., S. Fox, K. V., Proffitt, D. R., & Downs, J. H. III. (2002). *Imagined rotations of the self: an fMRI study. Department of Psychology*, Neuroimaging Research Laboratory, University of Virginia, Charlottesville, VA. http://www.psych.utah.edu/~sc4002/HBM99_Rotation.pdf

Erickson, M. H. & Rossi, E. L. (Ed.) *The collected papers of Milton H. Erickson on hypnosis: vol.* IV. Innovative hypnotherapy. NY: Irvington. 1980.

Erickson, M. H. (1954). Pseudo-Orientation in time as an hypno-therapeutic procedure. *Journal of Clinical Experimental Hypnosis, 2,* 261-283. In Milton Erickson & E. L. Rossi (Ed.) *The collected papers of Milton H. Erickson on hypnosis: vol. IV. Innovative hypnotherapy.* NY: Irvington. 1980.

References

Feil, J., Sheppard, D., Fitzgerald, P. B., Yücelc M., Lubman, D. I., & Bradshaw, J. L. (2010). Addiction, compulsive drug seeking, and the role of frontostriatal mechanisms in regulating inhibitory control. *Neuroscience and Biobehavioral Reviews 35*, 248–275.

Fidler, J. (1982). The holistic paradigm and general systems theory in *General systems theory and the psychological sciences*, vol.1, Gray, Fidler & Battista (Ed.)Seaside, California: General Systems Press.

Franken, I. H. A. (2003). Drug craving and addiction: integrating psychological and neuropsychopharmacological approaches. *Progress in Neuro-Psychopharmacology and Biological Psychiatry, 27*(4), 563-579. doi:10.1016/S0278-5846(03)00081-2.

Gendlin, E. (1981). *Focusing*. NY: Bantam.

Gray, R. M. (1996). *Archetypal explorations*. London: Routledge.

Gray, R. M. (1997a). Ericksonian approaches to the ego-self axis: Establishing futurity and a sense of self in addictive clients. Seminar: *Innovative Approaches to the Treatment of Substance Abuse for the Twenty First Century*. St. Francis College, Brooklyn, NY.

Gray, R. M. (1997b). Addiction And The Nature Of Meaning: Reframing In Substance Abuse Treatment. Seminar: *Innovative Approaches to the Treatment of Substance Abuse for the Twenty First Century*. St. Francis College, Brooklyn, NY.

Gray, R. M. (2001). Addictions and the Self: A self-enhancement model for drug treatment in the criminal justice system. *The Journal of Social Work Practice in the Addictions. 2(1)*.

Gray, R. M. (2002). The Brooklyn Program: Innovative approaches to substance abuse treatment. *Federal Probation Quarterly 66(3)*, December 2002.

Gray, R. M. (2003). The Brooklyn Program: Cognitive applications of the physiological correlates of spiritual experience. This paper was originally presented at *the Dr. Lonnie E. Mitchell National HBCU Substance Abuse Conference*, sponsored by Howard University, on April 2, 2003.

Grinder, J. & Delozier, J. (1987). *Turtles all the way down: Prerequisites to personal genius*. Scotts Valley, CA: Grinder Delozier Associates.

Gu, H., Salmeron, B. J., Ross T. J., Geng, X., Zhan, W., Stein, E., & Yang, Y. (2010). Mesocorticolimbic circuits are impaired in chronic cocaine users as demonstrated by resting-state functional connectivity. *NeuroImage 53*(2): 593-601.

Haley, J. (1973). *Uncommon therapy*. NY: W. W. Norton.

Hammond, D. C. (1990). *Handbook of hypnotic suggestions and metaphors*. NY: W.W. Norton.

Henderson, J. (1984). *Cultural attitudes in psychological perspective*. Toronto: Inner City Pub.

Hillman, J. (1975/1977). *Revisioning psychology*. NY: Harper Colophon.

Hillman, J. (1996). *The soul's code: In search of character and calling*. NY: Random House.

Huffman, K., Vernoy, M., & Vernoy, J. (1999). *Psychology in action*. NY: John Wiley & Sons.

Hull, C. L. (1933). *Hypnosis and suggestibility*. Englewood Clifts, NJ: Appleton Century Crofts.

Iacoboni M., Woods, R. P., Brass, M., et al. (1999). Cortical mechanisms of human imitation. *Science. 286*: 2526-2528. Reported at http://www.neurologyreviews.com/jan01/nr_jan01_emotion.html

James, T., & Woodsmall, W. (1988). *Timeline therapy and the basis of personality.* Cupertino, CA: Meta Publications.

Jung, C. G. (1966). *The Practice of psychotherapy (CW16).* Princeton: Princeton Univ. Press.

Jung, C. G. (1967). *Symbols of transformation* (CW5). Princeton: Princeton Univ. Press.

Jung, C. G. (1968). *Psychology and alchemy* (CW12). Princeton: Princeton Univ. Press.

Jung, C. G. (1968a). *Alchemical studies* (CW13). Princeton: Princeton Univ. Press.

Jung, C. G. (1977). *Mysterium conjunctionis* (CW14). Princeton: Princeton Univ. Press.

Jung, C. G. (1979a). *The archetypes of the collective unconscious* (CW9i). Princeton: Princeton Univ. Press.

Jung, C. G. (1979b). *Aion: Researches into the phenomenology of the Self* (CW9ii). Princeton: Princeton Univ. Press.

Kringelbach, M. L. (2005). The human orbitofrontal cortex: Linking reward to hedonic experience. *Nature Reviews: Neuroscience, 6*, September 2005, P. 691.

Kringelbach, M. L., & Berridge, K. C. (2009). Towards a functional neuroanatomy of pleasure and happiness. *Trends in Cognitive Sciences, 13*(11), 479-487. doi:10.1016/j.tics.2009.08.006

Langer, E. J. (1989) *Mindfulness.* NY: Addison-Wesley.

Laski, M. (1961). *Ecstasy in secular and religious experiences.* NY: Jeremy Tarcher.

LeDoux, J. (1995). Emotion: clues from the brain. *Annual Review of Psychology 46*, pp. 209-235.

LeDoux, J. (1998). *The Emotional brain.* NY: Touchstone.

LeDoux, J. (2002). *The synaptic brain.* NY: Viking Penguin.

Li, C.-s. R. & Sinha, R. (2008). Inhibitory control and emotional stress regulation: Neuroimaging evidence for frontal-limbic dysfunction in psycho-stimulant addiction. *Neuroscience & Biobehavioral Reviews 32*(3): 581-597.

Linden, A., & Perutz, K. (1998). *Mindworks: NLP tools for building a better life.* NY: Berkley Publishing Group.

Maslow, A. (1970). *Religions, values, and peak experiences.* NY: The Viking Press.

Miller, G. (1956). The magical number seven, plus or minus two. *The Psychological Review*, 63, 81-97.

Miller, G. A., Galanter, E., & Pribram, K. H. (1960). *Plans and the structure of behavior.* NY: Holt, Rinehart & Winston.

Miller, S. D., & Berg, I. K. (1995). *The miracle method: A radically new approach to problem drinking.* NY: Norton.

Miller, W. R., Zweben, A., DiClemente, C. C., & Rychtarik, R. G. (1995). *Motivational enhancement therapy manual; A clinical guide for therapists treating individuals with alcohol abuse and dependence.* Rockville, MD: NIDA.

Morris, R. G. M. (2006). Elements of a neurobiological theory of hippocampal function: The role of synaptic plasticity, synaptic tagging and schemas. *European Journal of Neuroscience 23*(11): 2829-2846.

Muhlberger, A. Neumann, R., Wieser, M., & Pauli, P. (2008). The impact of changes in spatial distance on emotional responses. *Emotion, 8*(2), 192–198.

References

Nyanaponika Thera & Bhikku Bodhi. (1993) *Abhidhamma Studies : Buddhist explorations of consciousness and time.* Somerville, MA: Wisdom Publications.

Ostrander, S., & Schroeder, L., & Ostrander, N. (1994). *Super learning 2000.* NY: Delacorte Press.

Overdurf, J. (2006, April). You never know how far a change will go ...Beyond goals. Pre-Conference workshop conducted at the 19th Annual Convention of the Canadian Association of NLP. Retrieved from http://johnoverdurf.typepad.com/ canlp/files/ canlpmanual.pdf

Peck, M. S. (1998). *The road less traveled: A new psychology of love, traditional values and spiritual growth (third ed.).* NY: Simon & Schuster.

Peele, S., Brodsky, A. & Arnold, M. (1991). *The truth about recovery and addiction.* NY: Simon and Schuster.

Pham, L. B., & Taylor, S. E.. (1999). From thought to action: Effects of process-versus outcome-based mental simulations on performance. *Personality and Social Psychology Bulletin, 25,* 250-260.

Piaget, J. (1970a). *Genetic epistemology* (Eleanore Duckworth, Trans.). NY: Columbia University Press.

Prochaska, J. O., Norcross, J. C., & DiClemente, C. C. (1994). *Changing for good.* NY: William Morrow.

Progoff, I. (1959). *Depth psychology and modern man.* NY: The Julian Press.

Rescorla, R. A. (1988). Pavlovian conditioning, *American Psychologist,* 43(3), pp. 151-160.

Robbins, Anthony. (1986). *Unlimited power.* NY: Fawcett Columbine.

Rossi, E. (2000). In search of a deep psychobiology of hypnosis: Visionary hypotheses for a new millennium. *American Journal of Clinical Hypnosis,* 42(3)/42(4), 178-207.

Rossi, E. L., & Cheek, D. B. (1995). *Mind-Body therapy.* NY: W. W. Norton.

Rossi, E. L. (1986). *The psychobiology of mind-body healing.* NY: W. W. Norton.

Samuels, A. (1985). *Jung and the post jungians.* NY: Routledge and Kegan Paul.

Schaeffer, H., & Martin, P. (1969). *Behavioral therapy.* NY: McGraw Hill.

Scheele, P. (1997). *Natural brilliance.* Lake Wayzata, MI: Learning Strategies Corp.

Schultz, W. (2002). Getting formal with dopamine and reward. *Neuron. 36(1).* September 26, 2002.

Simons, R., Detenber, B., Reiss, J., & Shults, C. (2000). Image motion and context: A between- and within-subjects comparison. *Psychophysiology, 37,* 706–710.

Simons, R., Detenber, B., Roedema, T., & Reiss, J. (1999).Emotion processing in three systems: The medium and the message. *Psychophysiology, 36,* 619–627.

Skinner, B. F. (1957). *Science and human behavior.* Garden City, NY: The Free Press.

Sparks, D. (1999). Conceptual issues related to the role of the superior colliculus in the control of gaze. *Current Opinion in Neurobiology,* 9:698–707.

Thorndike, E. L. (1911). Animal intelligence. Retrieved from http://psychclassics. yorku.ca/author.htm#t

Tulving, E. (2002). Episodic memory: from mind to brain. *Annual Review of Psychology 53*: 1-25. DOI: 10.1146/annurev.psych.53.100901.135114

Wake, L. (2010). *NLP principles in practice*. St.Albans, Hertfordshire, UK: Ecademy Press.

Watzlawick, P., Weakland, J. H., & Fisch, R. (1974). *Change: Principles of problem formation and problem resolution*, NY: W.W. Norton.

Wegner, D. M., Schneider, D. J., Carter, S., & White, T. (1987). Paradoxical effects of thought suppression. *Journal of Personality and Social Psychology, 53*, 5-13.

Wenger, W. (1979). *Beyond O.K.: Psychegenic tools relating to health of body and mind.* Gaithersburg, MD: Project Renaissance.

Wenger, W., & Poe, R. (1997). *The einstein factor.* Rocklin, CA: Prima Publishing.

Wolpe, J. (1958) *Psychotherapy by Reciprocal inhibition.* Stanford: Stanford University Press.

INDEX

20 minutes to two hours 87

abduction 106

Acupuncture 23

Addictive behaviors 6

altered states 20, 23, 26, 27, 31, 46, 67, 87, 90, 152, 165, 181, 185, 193, 218

anchored resources 9

anchoring ...... iii, 8, 9, 11, 19, 20, 21, 22, 23, 26, 48, 61, 64, 66, 74, 77, 81, 82, 85, 93, 95, 98, 102, 113, 114, 117, 146, 161, 191, 192, 198, 199, 201, 207, 215, 222, 223, 224, 235

anchors devoid of content 11

Andreas, Connirae 8

Andreas, Steve . xii, xiv, 7, 8, 12, 18, 19, 26, 28, 54, 57, 59, 71, 75, 77, 88, 120, 164, 239

archetype 8, 119

Attention deficits xi

autogenic training 121

Bandler, Richard xiv, 2, 3, 7, 11, 12, 18, 19, 20, 22, 24, 26, 28, 29, 30, 38, 43, 54, 57, 59, 60, 75, 76, 77, 78, 88, 89, 93, 103, 119, 120, 134, 164, 179, 182, 225, 239, 240

Bandura, Albert viii

basic states 20, 55, 70, 79, 151

Bateson, Gregory 106, 239

bicycle 3, 47, 49, 55, 70, 187

calling .... i, viii, ix, 105, 109, 116, 128, 191

Campbell, Joseph 128, 240

Casteneda, Carlos 76, 240

choice criteria 46

classical conditioning .... iii, 21, 88, 239

classically conditioned iii

clean, sober and legal. 20

color-blind participant 29

complex system 79

conditioned response ..19, 22, 61, 65, 80, 197, 221

conditioned stimuli ......iii, xi, 19, 23, 59, 62, 63, 79

Conditioning 67, 68

conscious choice 17, 20, 48

content-free.. 53, 62

Core Transformations................................ 8, 12

craving...4, 5, 7, 13, 241

criteria for choosing remembered states 27

D'Aquili and Newberg x

Damasio, Antonio 7, 12, 18, 20, 37, 43, 61, 75, 78, 84, 85, 121, 134, 239, 240

Daniel Schacter and Jerome Singer 23

Davidson, Richard 10, 13, 240

DaVinci, Leonardo 75

deep Self ....iv, v, 8, 9, 121, 144, 201

delayed conditioning...................................21

dials and knobs 76

difference that makes the difference...................75

Dilts, Robert ..xiv, 6, 13, 18, 19, 22, 26, 28, 54, 59, 71, 75, 77, 78, 88, 93, 103, 119, 134, 142, 143, 144, 145, 147, 148, 164, 240

disassociated states...................................26

disease concept...................................2

dizziness...................................82

dopamine............................4, 5, 6, 9, 10, 11

ecstasy.............9, 11, 17, 28, 30, 34, 52, 61, 73, 155

Einstein, Albert 138, 140

emergent property...................................vii, 23, 115

endogenous states...................................25

enhance memory...................................27

falling in love...................................7

feared stimulus...................................18

feeling tone...................................26, 48

felt experiences...................................iii, 9, 62, 71, 75

fine-grained detail...................................18

five simple gestures...................................22, 67

flavor...................................30, 34, 50, 52, 155

fMRI 122, 240

Fractionation...................................60, 76

Future-Pacing...................................88

Gendlin, Eugene...................................x, 241

general systems theory...................................ii

generalization 2, 7, 20, 23, 24, 60, 87, 88, 169, 194, 206, 218, 225

Generalization...................................88

generative...................................8, 120, 121, 213

gently turn...................34, 49, 52, 56, 62, 65, 155

Grinder, John .... xiv, 2, 3, 7, 12, 18, 19, 22, 24, 26, 28, 43, 54, 57, 59, 60, 75, 78, 88, 89, 93, 103, 119, 120, 134, 135, 164, 182, 225, 239, 240, 241

habituation...................................4, 5

hierarchies...................................2, 6, 10, 18

hierarchy of salience...................................9

Hillman, James 8, 14, 105, 109, 111, 118, 119, 241

hope .... xi, 11, 27, 41, 60, 146, 171, 182, 214

idiosyncratic...................................45

illegal, immoral ... objectionable states 28, 31, 152

ImageStreaming.. xiv, 101, 137, 138, 139, 140, 142

imaginary hands.. 30, 33, 34, 41, 43, 51, 52, 56, 76, 154, 155, 179

imagined representation 88

immoral or otherwise forbidden 20

incentive salience 4, 5, 6, 14

independent, self-maintaining behavioral networks 11

individuation .... viii, 17, 84, 102, 105

inhibitory deficits 7

instructional tone 29

Jung, C. G. .... iv, viii, ix, x, 8, 9, 14, 17, 84, 85, 86, 102, 105, 116, 119, 142, 242, 243

Jungian complexes 8

kinesthetic anchors 61

Law of Requisite Variety 90

LeDoux, Joseph 10, 14, 43, 84, 86, 121, 242

Maslow, Abraham .. viii, ix, x, 8, 14, 17, 19, 84, 86, 102, 105, 108, 113, 135, 242

meditation ..... x, xi, 23, 65, 68, 83, 87, 93, 127, 139, 146, 163, 165, 169, 177, 181, 185, 187, 203, 208, 218, 224

Memory Enhancement 43

Mickey Mouse 43

midbrain 4, 5, 6, 9, 10

midbrain dopamine tract 4

Miller's Magic Number 11

Minnesota Model 2

mnemonic 62, 196

more highly valued 7, 11

Morris, Richard 106

Motivation 120, 121, 134, 135

Moving stimuli 18

multiple contexts .... iv, 6, 9, 87, 88, 97

multiple sensory systems 19, 37

Neither content nor context 79

neuroscience sconfirming NLP 9

Nominalizations 2

NOW anchor v, 102, 106, 107, 114, 115, 119, 121, 202, 203, 208

Objections 48

onset of the state 76

open a window 49, 109

operationalization .... x, 119

outframe 7, 8, 9, 114

Pavlovian conditioning 18, 59, 243

peak experiences .... x, 102, 105, 108, 163

Peele, Stanton 3

personal direction . i, ix, x, 7, 8, 105, 109, 114, 115, 120, 175, 213, 223

preference hierarchy 6, 7, 10

presuppositions .. i, ii, vii, ix, xi, 26, 45, 54, 79, 113, 116, 119, 120, 217

Prochaska i, iv, viii, 10, 14, 121, 122, 137, 224, 243

Pseudo-Orientation in Time 120, 121, 134, 143, 240

Pseudo-Orientations in Time 88, 169

pursued for their own sake 10

Reframing 7, 12, 19, 110, 117, 148, 239, 241

reinforcers 5

Resistance Destroyer 20

resource states 19, 22, 23, 26, 45, 46, 50, 53, 59, 74, 84, 88, 106, 152, 170, 215, 223, 224, 225

Response Generalization 20

road rage 9

rules for choosing memories 55

rush of onset 30, 33, 51, 154

Russian Olympic teams 122

schedule of reinforcement 5

self-actualization .... viii, ix, 17, 84, 102

Self-Efficacy .... viii, 20, 45, 97, 98, 134, 139, 239

Self-esteem .... ix, 24

Sensory Enhancement 43

short term memory .. 11, 19, 26, 30, 34, 37, 52, 65, 155

signs of altered consciousness 53

single event 48

six step reframe 7, 8

six things 22

smart outcomes v

specific instance 48

spiritual .... v, x, 5, 7, 13, 17, 98, 103, 108, 122, 123, 125, 128, 130, 131, 175, 179, 181, 214, 224, 225, 241

spiritual experience 17

stacking anchors .... iv, 23, 113

Stages of Change .... i, ii, 10, 121, 122

Sternman, Shelly 7

stick shift .... iv, 47, 49, 55, 70, 187

strong principle of change 122

submodality ... iv, 20, 27, 28, 29, 31, 32, 35, 37, 38, 39, 40, 43, 50, 73, 75, 77, 152, 153, 177, 185

Submodality analysis iii

substances of abuse 4, 6

suggestions ..... xii, xiv, 7, 29, 49, 62, 182, 202, 203, 208, 222

Super Learning 77

Synesthesias 62

syntax .... iii, 18, 119, 120, 121

system principles 79

systems theory .... vii, 12, 240

time line 143, 144, 145, 146, 147, 148, 171

Timing of the anchor 62

transcendence 3, 5, 9, 68

transcendent states iv

Transcendental Meditation 23

transformative outcomes .... iii, 121

Transforming Emotions 43

triggers .... iii, vii, 22, 62

References

validated visual submodalities 18

Viet Nam 3

wanting something that is more important 10

well-formed outcomes .... iii, 122

Wenger, Win .... xiv, 103, 137, 140, 142, 164, 244

wholeness .... vii, 8, 26, 84, 105, 119

Yoga 23, 87, 179

Printed in Great Britain
by Amazon

60781192R00156